To Gordon

In the hope that you might find parts of this distracting on one of your many flights!

Best wishes

Ray

November 1999.

Agricultural Extension and Rural Development
Edited by Ray Ison and David Russell

While cultural traditions are important, there is a need to challenge, as well as to respect, those traditions which have given rise to a particular rural research and development system. The authors examine the broad themes of 'knowledge transfer' and the organisation and conduct of research and development from a 'second-order' systems perspective. The discussion is based on the situation operating in the rangelands of Australia, where the need to adopt a novel aproach arose because the existing practice of agricultural extension did not meet the needs of the farming community there, and the research funding bodies were dissatisfied with the return on their investment. The ways in which the relationship between the rural community and researchers might be more effectively managed are discussed. The issues addressed have relevance in a wider context and this book will therefore be of importance to any professional involved in agricultural management and policy making.

RAY ISON is Professor of Systems and Director of the Post-Graduate programme in Environmental Decision Making at the Open University.

DAVID RUSSELL is Associate Professor in the Faculty of Social Inquiry and the Head of the Centre of Research in Social Ecology at the University of Western Sydney.

Agricultural Extension and Rural Development: Breaking Out of Traditions

A second-order systems perspective

Edited by
Raymond L. Ison
The Open University, UK

and David B. Russell
University of Western Sydney, Australia

PUBLISHED BY THE PRESS SYNDICATE OF THE UNIVERSITY OF CAMBRIDGE
The Pitt Building, Trumpington Street, Cambridge, United Kingdom

CAMBRIDGE UNIVERSITY PRESS
The Edinburgh Building, Cambridge CB2 2RU, UK   http://www.cup.cam.ac.uk
40 West 20th Street, New York, NY 10011-4211, USA   http://www.cup.org
10 Stamford Road, Oakleigh, Melbourne 3166, Australia
Ruiz de Alarcón 13, 28014 Madrid, Spain

© Cambridge University Press 2000

This book is in copyright. Subject to statutory exception
and to the provisions of relevant collective licensing agreements,
no reproduction of any part may take place without
the written permission of Cambridge University Press.

First published 2000

Printed in the United Kingdom at the University Press, Cambridge

Typeset in Utopia 9/13 pt [VN]

A catalogue record for this book is available from the British Library

Library of Congress Cataloguing in Publication data

Agricultural extension and rural development: breaking out of traditions /
edited by Raymond L. Ison & David B. Russell
    p.   cm.
  Includes index.
  ISBN 0 521 64201 9 (hc.)
  1. Agricultural extension work – Australia – New South Wales.
2. Rural development – Australia – New South Wales.   3. Pastoral
systems – Australia – New South Wales.   4. Rangelands – Australia – New
South Wales.   I. Ison, R.L.   II. Russell, David B.
S544.5.A8A38   1999
630'.71'5 – dc21   98 51725 CIP

ISBN 0 521 64201 9 hardback

Contents

List of contributors   vii
About the authors   viii
Acknowledgements   x

Part I	**Breaking Out of Traditions** 1
1	The research–development relationship in rural communities: an opportunity for contextual science *David B. Russell and Raymond L. Ison* 10
2	The human quest for understanding and agreement *Lloyd Fell and David B. Russell* 32
3	Technology: transforming grazier experience *Raymond L. Ison* 52
Part II	**Historical Patterns, Technological Lineages and the Emergence of Institutionalised Research and Development** 77
4	From theodolite to satellite: land, technology and power in the Western Division of NSW *Adrian Mackenzie* 80
5	Experience, tradition and service? Institutionalised R&D in the rangelands *Raymond L. Ison* 103
Part III	**A Design for Second-order Research and Development** 133
6	*Enthusiasm:* developing critical action for second-order R&D *David B. Russell and Raymond L. Ison* 136
7	Co-researching: braiding theory and practice for research with people *Lynn Webber* 161
8	The graziers' story *Danielle Dignam and Philippa Major* 189

Part IV	**Limitations and Possibilities for Research and Development Design** 205
9	Designing R&D systems for mutual benefit *David B. Russell and Raymond L. Ison* 208
	Appendix 219
	Glossary 223
	Index 232

Contributors

Danielle Dignam
c/o Wendy Stamp
AWT Business Services
PO Box A239
SYDNEY SOUTH
NSW 1235
Australia

Lloyd Fell
NSW Agriculture
University of New England
ARMIDALE
NSW 2351
Australia

Raymond L. Ison
Systems Discipline
Centre for Complexity
and Change
The Open University
MILTON KEYNES
MK7 6AA
UK

Adrian Mackenzie
General Philosophy
Main Quad, A14
University of Sydney
NSW 2006
Australia

Phillipa Major
City West Development Corporation
PO Box 244
PYRMONT
NSW 2009
Australia

David B. Russell
School of Social Ecology
University of Western Sydney
  (Hawkesbury)
RICHMOND
NSW 2753
Australia

Lynn Webber
Community Relations Division
NSW National Parks & Wildlife Service
43 Bridge Street (PO Box 1967)
HURSTVILLE
NSW 2220
Australia

About the authors

Ray Ison and David Russell began their collaboration in 1986 whilst both were at the then Hawkesbury Agricultural College, Richmond, Australia (now University of Western Sydney (Hawkesbury)). The starting point was a seminar presented by David, which led to intense debate about objectivity and the nature of reality. This seminar triggered a conversation that is on-going. Both David and Ray continue to research questions that have arisen from the experiences reported in this book.

Ray Ison has degrees in agricultural science. He is currently Professor of Systems and Director of the Open University Post-graduate programme in Environmental Decision Making. He was Head of the Department of Systems (now the Systems Discipline in the Centre for Complexity and Change) at the The Open University in the UK from 1995 to 1998. Whilst engaged in this study he was a Senior Lecturer in Grassland Systems at the University of Sydney.

David Russell has science and psychology degrees. Following the study on which this book is based he was Director of the Centre for the Social Ecology of Water and Waste at Hawkesbury. David has been a major contributor to the development of Social Ecology at UWS–Hawkesbury where he is now Associate Professor in the Faculty of Social Inquiry. He is currently the Head of the Centre of Research in Social Ecology, which has as it aim the pursuit of postgraduate research that integrates the personal, the social and the ecological. The Centre brings together and supports research in the areas of psychology, spirituality, leadership, environment, and education.

Danielle Dignam is a graduate in systems agriculture from the University of Western Sydney (Hawkesbury). She has worked with pastoralists in the semi-arid region of NSW in a number of capacities including a period as a community facilitator with the NSW Conservation and Land Management Department (CaLM). She is currently living in Pakistan.

Lloyd Fell studied Rural Science in Australia completing PhD studies on animal physiology (brain hormones) at the University of Melbourne in 1973. He has conducted research in animal science in Australia and New Zealand since 1962. Currently a Senior Research Scientist with NSW Agriculture located at the University of New England, he is engaged in research on animal behaviour, stress, health and welfare of farm animals. His interest in issues of human cognition and communication grew out of his research on stress and was kindled particularly by a long-standing personal association with the University of Western Sydney, Hawkesbury and intermittent personal contact with Dr Humberto Maturana from the University of Chile.

Philippa Major is an educator and writer now working with the newly established Sydney Harbour Foreshore Society. Her recent publications include *Doors were always open. Recollections of Pyrmont and Ultimo* (1997), a chapter in *For a Common Cause. Case Studies in Communities and Environmental Change* (1996). She has an ever-increasing interest in how to support and communicate with communities through times of change.

Adrian Mackenzie is a postdoctoral research fellow at the School of Philosophy, University of Sydney, studying time, technology and embodiment. In dialogue with contemporary philosophical theory, and social studies of science and technology, his research focuses on how cultures make sense of technological change. His current interest lies in developing a critical account of the speed of that change in terms of the complexities of human–technical interactions.

Lynn Webber is currently Manager of Community Programs & Consultation at the NSW National Parks & Wildlife Service, a state government agency responsible for working with the community for the conservation of nature, Aboriginal heritage and historic heritage in New South Wales, Australia. Areas of specialist interest include action research/learning, community consultation design and facilitation, neighbour relations and strategic communications in natural resource management. In addition to work on the CARR project team and doctoral research in western NSW, Lynn has undertaken research and programme implementation in the Solomon Islands and a range of rural and urban contexts in New South Wales.

Acknowledgements

This book is based on eight years of active collaboration and three years of intense research in which we have attempted to develop alternative modes of doing research and development (R&D) with pastoralists in far western NSW, Australia. We experienced our research as challenging work. It was challenging emotionally, professionally, conceptually, institutionally and socially. To sustain such work requires nurturing and sustenance of those who participate because collaboration is hard work if it is genuine collaboration. It requires mutual respect and the valuing of difference – of leaving space for the other. It is an emotional process and needs to be recognised as such. When writing after a period which so engaged our emotions it is often difficult to capture the richness and intensity of those moments. The words of Elizabeth Jane Howard in her novel *the Long View*[1] come to mind: 'She discovered, much later, that any emotion, if it be sufficiently strong, is elusive, not accurately memorable – that only the small practical fringe – the attendant commonplaces remain vividly in the mind – and that the memory of them only serves to conceal the core essence of violently felt experience'. We raise this here because our book is about breaking out of a research and development tradition in which emotion, or emotioning, is considering to have no part to play.

That we have been able to complete this book in a period of considerable personal upheaval is, we believe, the product of our relationships with a number of people. We would like to acknowledge these relationships because in the network of conversation that they realise, we have been granted considerable knowledge. We thank the former Australian Wool Research & Development Corporation (now The Woolmark Company), which supported this research, and particularly the many graziers who made time for us and helped us to develop a lasting affection for their part of the world. RLI would particularly like to thank Cathy Humphreys for the lens she has provided into second-order cybernetics and feminist epistemology, especially from the family therapy literature. Our colleagues Rosalind Armson, Richard Clark, Greg Curran, Garry Hammond, Cathy Humphreys, Ross Humphreys, Janice Jiggins, Stephany Kersten, Bob Macadam, David

---

1 / Howard, Elizabeth Jane (1956). *The Long View*. Perennial Library, Harper & Row, New York.

McClintock, Craig Pearson, Pat Shah, Ruth Williams and Susan Wyndham have helped us in innumerable ways for which we are grateful. The book would not have come to fruition without the support of Alan Crowden, Maria Murphy and Katrina Halliday of Cambridge University Press. We and our co-authors however accept final responsibility for what we have written.

Part I
Breaking Out of Traditions

Traditions are very important to a culture because they embed what has, over time, been judged to be useful practice. The risk for any culture is that a tradition can become a blind spot when it evolves into practice lacking any manner of critical reflection being connected to it. When a society stops looking back and no longer appraises the value of a set of practices it quickly becomes blind to the relevance of its origins, the circumstances which were current at the time, and which triggered the practice into existence. The upshot is that there are no longer processes which foster the ongoing modification of the practice as a result of what we experience in daily living. The effects of blind spots can be observed at the level of the individual, the group, the organisation, the nation or culture and in the metaphors and discourses in which we are immersed. Such has been the case with 'research' and 'development' in rural communities. What began as a wonderful idea has evolved into blind practice as a consequence of the loss of connectedness with its context, the very connectedness that gave meaning and thus relevance to its existence in the first place.

What follows is a critical account of a systemic learning and researching approach to rural research and development (R&D). This approach arose from the need both to respect and to challenge the traditions which had given rise to a particular rural 'research and development system' in the semi-arid rangelands of New South Wales, Australia. The issues which this approach addresses have relevance beyond this specific context as exemplified in the works of Robert Chambers (Chambers 1993, 1996) and others (e.g. Pretty 1995; Roling 1997; Roling and Wagemakers 1998). This is a story of a systemic action research project (Table I.1) that sought to appreciate how the relationship between the rural community and the community of experts might be differently, and hopefully more fruitfully, managed.

The experiences which gave rise to the research described here arose out of a shared concern that the existing practice of rural development, or agricultural extension, was not meeting the espoused needs of some of the key stakeholders involved. Specifically, the majority of the rural community were not experiencing the expected benefits and the funding bodies, particularly those in the public domain, were dissatisfied with the return on their investment. As is often the case, this pragmatic concern was matched by a keenly felt intellectual concern, namely, that the conceptual under-

Table I.1
The contrast between traditional action research (a first-order tradition) and systemic action research (a second-order tradition)

Traditional action research	Systemic action research
The espoused role of the researcher is that of participant–observer. In practice, however, the researcher remains 'outside' the system being studied.	The espoused role and the action of the researcher is very much part of an interacting ecology of systems. How the researcher perceives the situation is critical to the system being studied. The role is that of participant–conceptualiser.
Ethics and values are not addressed as a central theme. They are not integrated into the change process; the researcher takes an 'objective' stance.	Ethics are perceived as being multi-levelled as are the levels of systems themselves. What might be 'good' at one level might be 'bad' at another. Responsibility replaces 'objectivity' in a whole systems ethic!
The system being studied is seen as distinct from its environment. While it is spoken of in 'open system' terms, intervention is performed as though it were a 'closed system'.	It is the interaction of the system with its context (its environment) that is the main focus of exploration and change.
Perception and action are based on a belief in a 'real world'; a world of discrete entities that have meaning in and of themselves.	Perception and action are based on one's experiences of the world. Especially on the experience of patterns that connect entities and the meaning generated by viewing events in their contexts.

(Source: After Russell, 1986).

pinnings upon which practice was based were faulty. They were faulty because they simply did not work! Theory that does not nurture useful practice is not useful theory.

The principle notions that were the driving ideas behind the movement for externally funded rural R&D were that 'best' knowledge and practice can be clearly articulated and that it can be effectively disseminated via a process of education. The implication is that the entire community will thus benefit through improved sustainability of the enterprises, increased production, and higher standards of living. Lived experience has confirmed that there is much that is obviously valid and useful in these principles. It is as though they worked but only up to a point. Over a period of forty years considerable effort was expended in an attempt to make the system more effective. Strategies included field-days, farmer groups and the involvement of farmer representatives on decision-making bodies, all of which resulted in little or no obvious improvement. Such was the general disillusionment of government agencies that this particular study had its origins in a political climate characterised by the dominant theme of: *any further expenditure on rural extension or rural development would be a waste of money*. This political climate was not confined to Australia, as examples in Chapter 1 demonstrate.

The conclusion of the research team prior to launching this work was that the existing system had gone as far as it could go and that 'more of the same' would not be of much use. Change which is 'more of the same' we shall call first-order change. The question that was left begging from an earlier 'critical review' of the overall situation was: perhaps we cannot see what the problem is because we cannot identify our own blind spots (Russell *et al.*, 1989). In technical terms, it is only when we step out of the system that we can begin to see the system from another perspective or from another level. The implication is that the other perspective or other level has a different rationale or basis for its existence. Change from this perspective we describe as second-order change. It is change which modifies the whole system.

The dominant tradition which gave rise to this situation we shall describe as the first-order tradition. This has been very powerful because of the manner in which it has shaped the actions of individuals, organisations and their structures, technology and the very language chosen to make sense of doing rural research and development. For this reason we will use the term R&D as a noun to break away from the traditions typically associated with 'research' and 'development'. We do so because models of understanding of 'research' and 'development', as with all models of understanding, grow out of a *tradition* – a network of prejudices (literally understood as a pre-understanding) that provide possible answers and strategies for action. Notions of what constitute both research and development are widely and firmly held in the community at large and by practitioners. This is why we choose instead to talk about R&D.

Having accepted that the traditional practice (including its underlying theoretical principles) had failed the pragmatic test: 'Does it work well-enough to keep doing it?', the challenge of designing and evaluating a more appropriate set of practices remained. What followed was three years of R&D by a group of 'systemic action researchers' concerned initially with exploring the context of their research and then linking themselves (by espousing a rationale of mutual benefit) to a community of wool growers (known as 'graziers') in the semi-arid rangelands of Australia.

The region in which this research was conducted is the Western Division of New South Wales, Australia. In 1990 there were 314 'establishments' in the NSW Western Division, a region totalling 32.5 million ha or 42 percent of the state of NSW, and 9.1 percent of the total area of semi-arid rangeland in Australia. The part of the Western Division in which our research was based covered 17 600 km^2 and included 45 properties, aggregated into 33 holdings (Figure I.1). Properties or 'stations' thus averaged *c.* 40 000 ha or 53 500 ha per family unit although these data mask the range in property

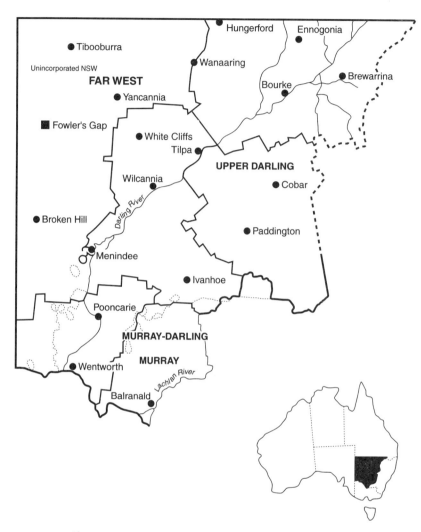

Figure I.1
A map of the Western Division of New South Wales and its location in Australia.

sizes. Estimates in the early 1990s suggested an economically viable property in this region needed to carry between 6000 and 9000 merino wool producing sheep. This figure is however highly variable depending on international commodity prices. The main agricultural product of the region is merino wool, described locally as 'middle micron' wool, indicating a fibre diameter in the range 22 to 26 microns. The Fowler's Gap area in which the research was centred, experienced a severe 'drought' – a natural and common phenomenon – for over two of the three years of this research project which, when combined with very low wool prices, produced a state of crisis for most graziers in this region. Road travel in the area was on unsealed roads made impassable by more than 10–15 mm of rainfall.

This R&D project resulted in an alternative design for rural development and all that is entailed in developing and sustaining mutually beneficial relationships between the so-called experts, the respective government bodies, and the primary producers themselves. Thus the book is a also a description of our attempts to break out of a first-order tradition and to design ways of doing rural R&D within a different tradition. This different tradition we shall call a second-order tradition. The major characteristic of this tradition is the continual attempt by researchers to be *aware* of their traditions of understanding.

As authors our collective concern has been with exploring, from a range of perspectives, the distinctions we make between first and second-order traditions. Heinz von Foerster (1992) highlights the profound implications of these distinctions when he says: 'Am I apart from the universe? That is, whenever I look am I looking through a peephole upon an unfolding universe [the first-order tradition]. Or: Am I part of the universe? That is, whenever I act, I am changing myself and the universe as well [the second-order tradition].' He goes on to say that 'Whenever I reflect upon these two alternatives, I am surprised again and again by the depth of the abyss that separates the two fundamentally different worlds that can be created by such a choice: Either to see myself as a citizen of an independent universe, whose regularities, rules and customs I may eventually discover, or to see myself as the participant in a conspiracy[2] whose customs, rules and regulations we are now inventing.' It is the response of a researcher or practitioner to this question that creates for us the distinction between action research (a first-order tradition) and systemic action research (a second-order tradition) (Table I.1). It is when what is done at any moment in privileging something and marginalising the other, without awareness, that one is operating in a first-order tradition. It is important to emphasise that both first and second-order traditions are modes of *doing* R&D, not labels for 'a tradition'.

An attempt to appreciate, or explore, one's context is one means to break out of a first-order tradition. Ison and Blackmore (1997) point out that an approach to dealing with complexity is to stand back from the apparent

---

2 / Von Foerster is of course using conspiracy here in the sense that has now been almost lost – to act in combination or to contribute jointly to a result. From its etymological roots it might also be defined as to breathe together and to bring forth the spirit, in a metaphorical sense.

issue and explore the wider context before inviting stakeholders to participate in a process of formulating and reformulating problems and opportunities or 'systems of interest'. At issue here is whether one is aware that all R&D is sensitive to its initial starting conditions and its on-going mode of practice.

At the start of our research it was possible to recognise three potential streams of inquiry which we felt it would be necessary to pursue if we were to fully appreciate our context. These were explorations of: (i) the traditions which have given rise to our very conception of rangelands, rangeland management and rangeland science (Chapter 1); (ii) the traditions which give rise to the meanings we give to human communication, and from this to information, knowledge and human understanding (Chapter 2) and (iii) the traditions which give rise to concerns about the lack of technology adoption and the common notions of technology transfer and the diffusion or trickle-down of innovations (Chapter 3). Chapters 1–3 explore what these traditions reveal and conceal and lay the theoretical groundwork for breaking out of existing traditions.

Technology has had a powerful influence in the design of what we distinguish and experience as rangelands. This is very evident from the technological trajectories within the 'Western Division' of NSW, as these semi-arid rangelands are sometimes called (Chapter 4). Not only do the rangelands arise out of traditions in understanding and technology but also the organisations which have been formed to research and develop the rangelands. This is revealed in Chapter 5, a case study based on the main organisations operating in the NSW rangelands over the three-year period of our project. What this story reveals, however, has more widespread relevance than just the semi-arid rangelands of NSW.

This book, and the research on which it is based, is constructed around the experience of the editors and the authors who were all involved in the project. It is the experiential history of practitioners and researchers which informs all action yet when this action is reported it is usually done so in a way that excludes this history from the conversation. We think it is important to bring this back into the conversation because it is one of the most important aspects of understanding or appreciating a context (Chapters 3 and 6). Recent R&D approaches, whether under the banner of 'farmer first', 'farming systems research' or whatever have focused our collective attention on the need to appreciate context in the R&D process. We suggest, however, that insufficient attention is still paid to the context of the researcher or development manager. By this we mean the traditions out of which they think and act.

A number of key concepts are at the core of our attempts to break out of the dominant, first-order tradition. Our central concern has been with the emotion of *enthusiasm*, and how this might be developed as both an organising concept and a methodology. This is explained in Chapter 6. Our research design for triggering enthusiasm was built around the experiential model of doing science proposed by the Chilean neurobiologist and epistemologist, Humberto Maturana (see Maturana and Varela, 1987). We know of no others who have used this model to conduct 'social research'. In Chapter 7 we describe in more detail how we went about our research to ellicit enthusiasms for R&D action. Our intention has been to avoid recipes to follow but rather to ground our work and enable an appreciation of the care and detail for process design that is required for this type of R&D.

As is most often the case there is often a disparity between design and realisation – any systemic action research over a three-year period almost inevitably will produce a rich and somewhat unfinished story. 'Rich' because of the contrasting needs of the respective groups and 'unfinished' because of the exploratory nature of the task. This richness can be conveyed in a number of ways, and as is increasingly the case, it is necessary to evaluate or judge actions from a number of perspectives. Different stakeholders will judge from different perspectives. The graziers who became our co-researchers based on their enthusiasms, describe their experiences of this way of researching in Chapter 8. For some, but not all graziers, we had designed a context where they were able to respond and ultimately to see themselves as 'researchers'.

Finally we are concerned with exploring what it means to break out of a tradition, what forms of rural R&D might be possible within this different tradition and how might we develop the skills to do this type of R&D (Chapter 9). A 'person specification' for a position in South Africa which combines the elements of what we see as being necessary to both build and utilise capacity for addressing rural R&D problems is shown in Figure I.2. This particular combination of abilities matches well what we envisage will be needed to move the R&D system more towards a second-order tradition whilst retaining the strengths of the first-order tradition.

Clearly there were strengths and limitations to our research, but what have we learnt from the experience and how has it informed what we have done since? What do we aspire to do in the future? We re-emphasise our claim for greater critical reflection on practice. As professionals, researchers, activists, facilitators, managers, academics and learners the challenge we face is to recognise and 'design' contexts which provide the capacity for effective response by stakeholders. However the question we must ask is how would we know the capacity for effective response when we see it? This is not a question of empowering the individual or enabling the individual to participate; for us it is the emergent relationship between

Figure I.2
Person wanted! A possible person specification for delivering R&D within a second-order tradition. (Courtesy of Ian Scoones and derived from a South African National Rural Development Forum advertisement in the Weekly Mail in June 1994 and a request from the Department of Land Affairs, November 1994.)

The Key Issue:

Building Capacity to Build Capacity

PERSON WANTED!

- Knowledge and experience of land, agriculture and rural development issues in South Africa
- Process skills and systems perspective on organisational development and capacity building
- Training, adult learning and group facilitation skills
- Negotation and conflict management skills
- Experience with participatory learning approaches in field and workshop settings (e.g. PRA)
- Experience with working in large bureaucratic public agencies and facilitating organisational change
- Personal authority and presence
- Willingness to travel extensively, especially to remote rural areas
- Fluency in several South African languages, English and Afrikaans

the enthusiasm of the individual and any consensus which is generated responsibly and accountably by a collective (Chapter 9). These are of a different level or order and thus do not represent a dualism, an either/or, but a duality, a unity (see Chapter 1).

In our research we learnt that enthusiasm was something that could be triggered, and that where there was enthusiasm there was action which was meaningful to that individual. We also learnt that processes which lead to consensus can get in the way of enthusiasm – there was loss of emotional energy for action. This is often experienced in relationships subject to repeated compromise. Based on this experience it would be easy to see enthusiasm and consensus as opposites, as belonging to the same logical level and forming a dualism. The logic behind this relationship or dialectic is negation. This we suggest is a trap. In contrast it is possible to see enthusiasm and consensus as belonging to two different levels such that one emerges from the other. The logic behind this dialectic is self-reference. This is exemplified by considering the pair predator/prey from ecology. They do not operate as opposites but generate a whole, a unity or an autonomous ecosystem where complementarity, stabilisation and survival are common values for both.

Thus we wish to ask more than just the question: How would we know the capacity for effective response when we see it? We wish to explore what a possible R&D system might be that retained some of the strengths of the first-order tradition but which explored and developed some of the opportunities presented by moving more of the overall R&D system towards the second-order tradition. The system we imagine would pay greater attention to project formulation – systems to express demand, to use a now common metaphor – systems of process consulting, novel systems of evaluation, and a rich array of co-researching activities. Any move in this direction would challenge many individuals and organisations, not least being higher education in which we are both engaged.

References

Chambers, R. (1993). *Challenging the Professions. Frontiers for Rural Development.* Intermediate Technology Press, London.

Chambers, R. (1996). *Whose Reality Counts?* Intermediate Technology Press, London.

Ison, R.L. and Blackmore, C.P. (1997) Exploring the context of environmental issues and formulating problems and opportunities. In *Environmental decision making: a systems approach, Block 2*, pp. 41–66. Open University, Milton Keynes.

Maturana, H. and Varela, F. (1987). *The Tree of Knowledge. The Biological Roots of Human Understanding.* New Science Library, Shambala Publications, Boston.

Pretty, J.N. (1995). *Regenerating Agriculture. Policies and Practices for Sustainablity and Self-reliance.* Earthscan, London.

Röling, N. (1997). The soft side of land. Socio-economic sustainability of land use systems. Proceedings Geo-information for sustainable land management conference, Enschede, the Netherlands 17–21 August 1997. *ITC Journal*, nos 3 and 4, 248–62.

Röling, N.G. and Wagemakers, M.A.E. eds (1998). *Facilitating Sustainable Agriculture. Participatory Learning and Adaptive Management in Times of Environmental Uncertainty.* Cambridge University Press, Cambridge.

Russell, D.B. (1986). *How we see the World Determines what we do in the World: Preparing the Ground for Action Research.* Mimeo. University of Western Sydney, Richmond.

Russell, D.B., Ison, R.L., Gamble, D.R. and Williams, R.K. (1989). *A Critical Review of Rural Extension Theory and Practice.* University of Western Sydney, Richmond.

von Foerster, H. (1992). Ethics and second-order cybernetics. *Cybernetics and Human Knowing*, **1**, 9–19.

1 The research–development relationship in rural communities: an opportunity for contextual science

David B. Russell and Raymond L. Ison

1.1 Introduction

This chapter argues for a contextual grounding for research and development (R&D) in rural communities. The history of science reveals many examples of how science has failed to recognise its context. So, what is context and how does one recognise it? It would be all too easy to answer these questions by simply adding social and political insights to the science equation. (What is necessary is that we look at the bigger picture!) Almost always, the bigger picture is nothing other than more of the same.

In this chapter we explore how our understanding of R&D is developed and how our understanding of 'change' is constructed. We are proposing what we believe to be a critical distinction based on the perceptions and actions of the researcher. In **first-order R&D**, which remains most common, the researcher remains *outside* the system being studied. The espoused stance by researchers is that of *objectivity* and while the system being studied is often spoken of in *open system* terms, intervention is performed as though it were a *closed system*. Perception and action by researchers and those who manage and maintain the R&D system are based on a belief in a *real world*; a world of discrete entities that have meaning in and of themselves.

In contrast to this tradition we stress the need for a **second-order R&D** in which the espoused role and action of the researcher is very much part of the interactions being studied. How the researcher perceives the situation is critical to the system being studied. *Responsibility* replaces objectivity as an ethic and perception and action are based on one's experiential world rather than on a belief in a single reality 'real' world. There are of course implications in any move towards a second-order R&D, not least of which are the forms of behaviour and organisation that might be required by, and for, a future cadre of 'researchers'. This is taken up specifically in Chapter 9, but much of the rest of the book is concerned with doing or moving towards second-order R&D.

1.1.1 *The global R&D system*

In his study of how scientists and engineers go about their work, Bruno Latour (1987) demonstrates with some simple statistics that those who call themselves scientists and engineers make up only a small proportion of the people interested in the generation of 'new knowledge' within the 'R&D

system'. It would seem that the number of scientists and engineers rarely exceeds 0.6% of the workforce, yet the practices which they largely initiate, give rise to technologies, metaphors, 'facts', and forms of organisation that affect profoundly the actions we take on a daily basis. In the US this was backed by an investment in 1988 of $139 billion in R&D (UNESCO, 1993) in a total worldwide investment exceeding $229 billion (Howells, 1990). The OECD maintains a database with 'rules of thumb', for nations to pursue as a guide to how much of their GNP should be invested in R&D. The R&D network is a powerful club.

Global R&D is growing (Howells, 1990). Most is conducted in the OECD countries, which is also where most of the 'researchers' are located (only 12.6% were in the developing world in 1973). As Janice Jiggins points out (Jiggins, 1993) an 'increasing number of countries in ... Africa, but also in Latin America, are facing the collapse of public sector research and extension.' Increasingly there is no effective institutional capacity for R&D in many of these countries.

Within the domain of natural resource R&D, in which we might include rural R&D, the World Bank, FAO, IFAD and the CGIAR[3] network have been the most active funders and supporters of R&D. Janice Jiggins (1993) reviewed the extent of funding committments by these agencies to R&D projects in which there has been an 'extension' or 'technology transfer element' (Table 1.1). She points out that in the R&D sector associated with rangelands and extensive livestock, only local and at most short-lived gains have been generated and that many of the R&D projects have had unintended negative consequences.

Ian Scoones (1995, p.3) claims that the 'last 30 years have seen the unremitting failure of livestock development projects across Africa. Millions of dollars have been spent with few obvious returns and not a little damage.' He notes that many donors and international agencies have abandoned the dry zone in their development efforts. He strikes a positive note, however, asking whether we should 'reconsider, and analyze in detail why the failure has been so consistent and what lessons can be learned from the convergence of recent ecological thinking, social science critiques and pastoralists' own practices?'

1.1.2 *Historical context*

As we arrive at the end of this twentieth century, it is useful to conceive of ourselves as living in the final days of near absolute faith in 'first-order' R&D.

---

[3] / FAO, Food and Agriculture Organization of the United Nations; IFAD, International Fund for Agriculture and Development; CGIAR, Consultative Group for International Agricultural Research.

Table 1.1
A summary of rural 'extension' related R&D expenditure by major funders

Agency	Period	Number of projects	Amount (US$)
World Bank	1965–89	500	2 billion
World Bank	1988–89	221 (extension related)	793 million
FAO	1985	977	57.4 million
IFAD	1978–88	191	320 million

The ethos and achievements of this period are characterised by disciplinary knowledge, a 'fix' mentality and the belief that the generation of 'new knowledge' is a good thing in itself. In rural-oriented R&D much attention has focused on exposing breakdown and attempting to fix it. Often the cry has been: More resources need to be deployed into researching the needs of rural industries! The development of this approach has had its own phases, all of which exemplify the fix mentality:

  (i) The 'problem' is seen as a mismatch between what is scientifically known and technically feasible, and what is current practice. The new technology is designed by research scientists and is then transferred to the end-users who put it into action to address the problem.
  (ii) Built into the belief of a technological solution is a conception of the benefits that could be derived from better farming systems or, in the case of rangelands, a return to the 'natural ecosystem' state, without consideration of who participates in defining 'better' nor how what is perceived as 'natural', by some, has come to be constructed.
  (iii) Social and political insights are specifically added to the R&D equation. (The declared purpose of the former International Livestock Centre for Africa [ILCA] would be an example of this multi-disciplinary approach to the development of models for range management.)

'Second-order' R&D challenges the first-order tradition, a tradition in which most of us are deeply immersed because of our cultural background and specifically because of our scientific training. We have labelled this tradition 'first-order' because of its emphasis on particular styles of consciously rationalised thought and action. Explicitly, it is a tradition based on a belief in an increasingly knowable world: a world which is capable of being understood without the need to take into account our actions as participants in creating that very world that we experience. There is a basic assumption that a fixed reality is 'out there' and that by applying rational understanding, we will increasingly gain accurate knowledge of its elements and the laws of its functioning. In addition, most often there is no distinction made between the possible understandings of material and biological phenomena (observable to the senses) and phenomena that are the products of the intellect (thoughts, beliefs, memories and the like).

We do not privilege first-order thinking with the widely held belief that it is the sole basis for being 'rational'. In questioning this there is no intention of fostering irrationality or fuzzy thinking, rather, along with Winograd and Flores (1987) our commitment is to developing a new ground of rationality – one that is as rigorous as the first-order tradition in its aspirations but that does not share the presuppositions underlying it.

At its simplest, the first-order view accepts the existence of an objective reality, made up of things bearing properties and entering into relations. We are actors in/on our 'environment'. Such has been the success and prestige of modern science that many accept it as the best framework available for understanding how we think and are intelligent.

1.1.3 *The origins of second-order R&D*

Developing out of this traditionally accepted paradigm is a much newer tradition that avoids being either objective or subjective. This tradition brings together understandings derived from the study of interpretation, the philosophical examination of the foundations of experience and action, and the 'new' *biology*, which provides an intellectual framework in which phenomena of interpretation arise as a necessary consequence of the structure of biological beings (see Chapter 2). All three intellectual streams have in common the questioning of our ability to objectify knowledge and thus see objects and events as being independent of the very act of observation. This new tradition avoids being either 'objective' (independent of the individual) or 'subjective' (particular to the individual). Our aim is not to replace scientific method but rather to show how our theoretical background might guide the design of research and development in the practical setting of the rangelands. By 'unpackaging' the presuppositions of the first-order interpretation of science, we become aware of its non-rational implications. This is especially the case in those most common of situations when there is no clear 'problem' to be solved, but a sense of irresolution that opens opportunities for action.

The region described as 'rangelands' provides some dramatic examples of first-order R&D and its unintended consequences.

1.2 Conceptual models of rangeland development

Examining the current practice of range management in any particular geographical context allows us to formulate the 'model of understanding' that informs those particular practices. The very strong emphasis on the production of beef, on commercial ranching, on the specialised stratification of the production process in breeding, on markets, and on processing facilities are characteristic of say North America and Australia. These characteristics are 'a reflection of an ideal of what pastoral development is about' (Sanford, 1983, p. 6) and have exercised a strong influence in much of the developing world. When we 'unpackage' the history of these developments we find that the American and Australian models originated 'in particular historical settings where the interests of the previous inhabitants of pastoral areas were not taken into account, where the (indigenous) species of domestic livestock of pastoral areas were not taken into account, where the species of domestic livestock on which pastoral development focused did not previously exist on a significant scale if at all, where the general economy as a whole was characterized by labour shortage rather than by surplus, and where a large and wealthy non-pastoral sector could be called on from time to time to provide the resources with which to rebuild a pastoral sector suffering from collapse' (*ibid*).

1.2.1 *The first-order tradition*

The first-order tradition is characterised by concerned intervention, the definition of clear goals, the 'naming' of the problem, and the proposal of a rational 'solution'. However, every model of understanding grows out of a *tradition* – a network of prejudices (literally understood as a pre-understanding) that provide possible answers and strategies for action. A 'tradition' here is taken to mean a pervasive, fundamental phenomenon that might be called a 'way of being.' A tradition is an intellectual background within which we interpret and act. In using the word 'tradition' we are emphasising the historicity of our way of thinking – the fact that we always exist within a pre-understanding determined by the history of our interactions with others who share the tradition (in Chapter 2 this definition will be expanded to incorporate what Maturana (1988) has termed our history of 'structural coupling').

An example is provided by exploring Le Houerou's (1989) work on the grazing land ecosystems of the African Sahel. It is possible to identify a number of themes that go to make up his 'way of thinking' and his way of constructing his working reality (his epistemology). First, there is a deeply felt concern for the ecology of the region. Second, there is a plea for the detailed and careful description of 'the philosophy and development objectives . . . and the strategy and means to attain the selected goals' (p. 239). Third, there is a clear statement of the core 'problem': 'adapting stocking rates to the sustained long-term productivity of the grazing ecosystem' (*ibid*). And finally, there is the proposed 'solution': 'responsible management . . . (which) involves fundamental land reform in terms of land tenure and ownership and water usufruct' (*ibid*). In order to appreciate this important contribution to the understanding of the Sahel we benefit greatly by looking at the tradition out of which it flows.

The first-order tradition can be depicted as a series of steps:

1. Characterise the situation in terms of identifiable objects with well-defined properties.
2. Find general rules that apply to situations in terms of those objects and properties.
3. Apply the rules logically to the situation of concern, drawing conclusions about what should be done. (From Winograd and Flores, 1987, p. 15.)

These steps are applied in a social context that encourages 'concerned' intervention, the definition of clear goals, the naming of the 'problem', and the proposal of a rational 'solution.' What is not encouraged is a debate about how the objects and properties were arrived at and how we come to know general rules, not to mention the issue of whose 'concern' is being attended to. Le Houerou's work raises all these first-order issues. It also represents an invitation to address the second-order issues which arise as we explore the intellectual context in which the first-order issues are embedded.

It is important to stress at this point that this exploration of the dominant tradition is designed to improve the application of good science and not to replace it. What is being proposed as this chapter unfolds is a contextual science for rural R&D that will evidence greater coherence with the expressed needs of the day-to-day lives of the people involved.

An inevitable experience of being embedded in any tradition is that we are not aware of the prejudices (pre-understandings) that shape our thinking and our action. This background of pre-understanding invisibly shapes what we choose to do and how we choose to do it. There is no neutral viewpoint from which we can see our beliefs as things, since we always operate within the framework they provide. This 'closed system', as it were, does not negate the importance of trying to gain greater understanding of our own assumptions so that we can expand our horizon. But it does preclude the possibility that such understanding will ever be objective or complete.

1.2.2 *When traditions of understanding collide*

A wonderful illustration of radically different 'frameworks of understanding' has been provided by Louise Fortmann's (1989) case study of fifty years of rangeland use in Botswana. Official policy consistently defined the major problem of the pastoral regions as overstocking leading to certain ecological disaster. The problem was clear, as was the technical solution (destocking). Local experience, on the other hand, defined the problem as too little land. The local solution was also very different: renting, or simply using an enormous concession of land previously given to a European mining company. The local experience was that the local range could and did carry an increased cattle population and that besides localised problems, the dire official predictions did not eventuate. While there is general agreement that the quality of the environment (as indicated by the quality of the grazing, the number of trees and the extent of erosion) is deteriorating, there was, for over fifty years, clearly no agreement on causes or solutions. Of particular significance for our argument is the story consistently told by both 'world views' and spanning such a long period; a story that shows how different and how unconnected traditions of understanding can be. What is more this is not an isolated example. Leach and Mearns (1996) review other examples from Africa of conventional wisdom which, on further study, may be deeply misleading. They point out that it is often in the interests of certain people and organisations to continue such myths. These examples are not confined to Africa or less developed countries (Pearson and Ison 1997; Chapters 3 and 4).

Stephen Sanford (1983) addressed this central issue of traditions of understanding in considerable detail when he talked of the 'Mainstream view' and what it entailed. This tradition was promoted by 'concerned' professionals (academics and officials in national and international organisations) and related to the belief that the world's rangelands were suffering severe and rapid desertification. As with any example of a first-order tradition, the problem is clearly defined, the solution is a technological one, and the 'barriers' to adopting the solution are placed fairly and squarely with the pastoral community: 'traditional economic and social systems, including systems of land tenure and the social institutions which accompany them' (*ibid*, p. 12). Along with Fortmann, he contrasts this 'Mainstream' view with the day-to-day experience of the pastoralists and the value, gained of generations of practical usage, of traditional systems. The lack of participation by pastoralists in the design and implementation of rangelands projects in the developing world has been a consistent criticism made in many published reviews of project effectiveness (Little, 1982; Ndagala, 1985; Gilles, 1985; Hunter, 1990; Scoones, 1995). Growing awareness of this situation has led some to adopt a new optimism (see Scoones, 1995) but it is still too early to tell if this optimism is warranted.

Range science with its twin goals of the protection of the environment through the concept of sustainable yield and the improvement of the productivity of ranges, had its origin in North America and its rapid adoption in Australia. Since range science and range management developed in North America, its approach was necessarily adapted to the social and ecological milieu of North American rangelands. A central feature of this history is that range management has evolved to meet the needs of a system based either on privately owned land or, as is largely the case in Australia, on land owned by the state and leased to individual livestock producers on a long-term basis, so that it is managed much as private property would be. So pervasive is this history, which constitutes this particular 'tradition of understanding', that it is difficult for those involved in it to see range management in any way other than their own way. This becomes very obvious when the privatisation of rangelands is considered to be a precondition for the protection of natural resources (Baden and Stroup,1977; Hopcraft, 1981). It is additionally apparent when the techniques of range management that have been developed in the West are applied, and have consistently failed in the less developed world (Gilles,1985; Lane and Moorehead, 1994; Lane, 1998).[4] The thought that they could possibly be effective in the first place is indicative of the continued blindness to seeing that such knowledge is socially constructed and is thus only applicable to its place of origin.

In a carefully constructed critique of the dominant paradigm of pastoral ecosystem dynamics, James Ellis and David Swift (1988) argued that the time was ripe to examine the paradigms which govern our thinking about African pastoral ecosystems. These authors readily acknowledge the 'social construction' of range science and specifically, the notion of an African pastoral ecosystem and what constitutes it. While their work is the result of a nine-year study in northern Kenya, the underlying principles of their research are equally applicable in any region of the world. The central idea that they hold up for critical scrutiny is the assumption that the African pastoral ecosystems are potentially stable (equilibrial) systems which become destabilised by overstocking and overgrazing (reflected in the work of Lamprey (1983) who argued that overstocking by pastoralists causes departures from natural ecosystem equilibrial conditions and range degradation). Their empirical results present the opposite view: pastoral systems that are non-equilibrial but persistent, with system dynamics affected more by abiotic than biotic

---

4 / It can be argued that much of this failure was due to bad technology and bad science, and no doubt some of it has been. Admission of this argument, however, means that advocates of traditional science need to be open to the same possibility (i.e. poor practice) when providing critiques of participatory research approaches. Our position is that because of the traditions of understanding in which we are (often unknowingly) immersed, the reasons for the many failures are more complex and profound than just good or bad science.

variables. Because 'Our view of the world, or our perceptions of any system, has a great deal of influence on how we go about dealing with that system' (p. 450), the conventional development practices are based on the assumption of equilibrial grazing systems and that destabilisation of these systems is due to overstocking and overgrazing by pastoralists. These practices have involved the establishment of group ranches, grazing block, or grazing associations which have not worked. Their conclusion is that conventional development practices are destabilising influences in ecosystems which are dominated by 'stochastic abiotic perturbations and which operate essentially as non-equilibrial ecosystems' (p. 458). This is a fascinating story as it so tellingly illustrates that we know the world only through our conceptual models of it, which themselves arise through our action-in-the-world.

Development interventions which arise, as they always must, from our model of the world, our tradition of understanding, and which do not flow from the traditional understanding of the pastoralist community, will always be 'development experiments' that will have unfortunate implications for the ecosystem and people on which they are performed. It was Martin Andrew's view, which he conveyed in his plenary paper for the Third International Rangelands Congress in New Delhi (1988), that many of the research and technical interventions that were reported during the course of the Congress were developed without understanding the behaviour and needs of the pastoral people. Sandford (1995) in his analysis of 'new directions' in pastoral development, based on the appreciation that they are non-equilibrial systems, confirms that many development schemes were misguided: 'the expensive and authoritarian ways of regulating livestock numbers, the dividing up of ranges into self-sufficient blocks and the creation of private ranches to bring these about proved to be inappropriate'. This raises the question of how much longer we need to be told that the 'mainstream view', 'the dominant paradigm', the 'top-down' approach, just does not work and is, in fact, detrimental to both the people and the ecosystem of which they are a part?

1.3 Technology 'transfer' or 'creation'

The first-order tradition in which we are immersed emphasises thought and its application (the generation of technology) as an independent activity. 'Knowledge' and 'applying knowledge' are the very language of R&D, a language that does not acknowledge its dependency on interpretation. The notion of 'information' as it is commonly used implies that an 'external world' is knowable in a way that is independent of the user of the language. In the first-order tradition, the information and the knowledge are 'out there' and one can collect more and more information about the external world and the greater the 'knowledge base', the greater the chances of useful technology and better interventions. The current trend to make technology 'user-friendly' is indicative of the questioning of the naive equation that more information equates with better results. While this questioning might not lead to a questioning of the theoretical paradigm itself, it will lead to the increased development of a technology designed 'to facilitate a dialogue of evolving understanding among a knowledgeable community' (Winograd and Flores, 1987, p. 76).

In a review of rural extension carried out by the authors (Russell *et al.*, 1989, 1991), it was concluded that the existing model of extension did not work well at all. It constituted neither good practice nor good theory. Promotion of innovative technology to the rural community has been based predominantly on the linear extension 'equation':

Research → knowledge → transfer → adoption → diffusion

A study of the effectiveness of this model showed that research results were adopted by only a specific minority of farmers and that for the majority, it was not a viable strategy for agricultural improvement. Experience of the deficiencies of this model in actual practice has led to the emergence of a very different conceptual system based on the idealised 'farmer-led' model (Chambers *et al.*, 1989). Despite the very real differences, both models incorporate current ways of thinking about and doing 'extension'. We think that it is time to abandon the term extension altogether because of what it has come to mean in practice and the network of faulty assumptions which are at its core.

As with range management, the term 'extension' arises from a particular tradition – from the North American land grant university model meaning 'to extend knowledge from a centre of learning to those in need of this knowledge'. Extension in practice has remained captive of this initial western conception despite differences culturally apparent in, say, the German '*beratung*' (to counsel or deliberate) and the French move from '*vulgarisation*' (to render popular) to '*développment agricole*' (involving the whole farming community).

1.3.1 *Information transfer*

The belief that knowledge could be 'transferable' has derived from the associated belief that 'communication' was the process of transmitting information. The media is convinced that we are now in the 'Information Age' so it is not surprising that the most widely used metaphor for the practice of extension is that of 'information transfer'. So embedded is this notion, so pervasive has been the obviousness of electronic communication, that challenging the appropriateness of continuing to use only this metaphor, is to risk being considered absurd. Risky or not, it must be done! The effectiveness of current practice continues to be judged and to be judged negatively (see Russell *et al.*, 1989, for a review of the literature and Scoones and Thompson, 1994, for a summary of emerging responses to this critique). Not only has the simple notion that knowledge can be transferred from one person to another, as if it were a case of one computer 'talking' to another, been shown not to work in practice, but biologically (as will be shown below and in Chapter 2), it is clearly not possible.

Shannon and Weaver (1949) were the first to use the model of electronic information transfer to refer to human communication. Simply put, they proposed that ideas were coded into signals, the messages, (by the sender) and then transmitted to another person (the receiver), who then decoded the message back into the original ideas. The root metaphor has had numerous elaborations in its application, such as those variously described as evidencing the *conduit* metaphor (Reddy, 1979) or the *hypodermic* metaphor. In the first instance, ideas were seen as being packaged into words so as to gain access to the original ideas. The hypodermic understanding was obvious when there was an intention to persuade the other to follow a certain course of action. The effective communicator could 'get under the skin' of the other if he or she could present the information 'persuasively'. David Sless (1986) has analysed a number of recent communication models showing how the basic 'information transfer' metaphor still dominates the thinking of many communication theorists. The prevalence of this established way of understanding communication, despite all the evidence to the contrary (see Sless, 1986 and Krippendorff, 1993 for reviews of the literature), shows how difficult it is to unearth a deeply embedded metaphor when it has taken root in the society's unconscious. The process of constructing more fitting metaphors will initially be awkward and cumbersome because we will inevitably have a foot in the old camp of *fixed reality*, a condition of the knowledge transfer idea, and a foot in the camp of *multiple realities*, the prerequisite for any new constructions.

1.3.2 *The biological basis of knowing*

In the language of the communication engineers, information is taken to mean 'instructing with knowledge'. What holds for engineering, in which communication systems are designed and structured with the intention of transferring information, does not hold for biological systems and in this discussion, we specifically mean human 'systems'.

Humans are structure-specified systems and cannot be instructed with knowledge by another living system (see Maturana & Varela, 1980, 1988). It is one's history of interactions and the closed self-generating structure of the human (autopoiesis) that determines what will happen and not the nature of the information. Often the observer acts *as if* there was a case of instruction by knowledge but this cannot be the case biologically.

The nervous system is a closed network of interacting neurons. The physiology of the nervous system, because it is a structure-determined system (systems in which all their changes are determined by their structure and in which all those changes are a result of their own dynamics or triggered by their interactions with their environment) cannot be usefully compared to a computer or 'information transfer' system. Biologically, there are no inputs to, or outputs from, the nervous system, nor does the nervous system 'process information'. There is no encoding or decoding in the nervous system nor does it 'receive' or 'process' messages or 'information' from the environment.

The implication that flows from the nervous system being a closed and structure-determined system is that there can be no instructive interactions between such systems and between any one system and its environment. What another human can do, and all that another can do, is trigger a response without any control over what that response might be. In no way can such a triggering determine the nature of the response. It is biologically impossible to instruct or determine an outcome with 'information'.

1.3.3 *Structural coupling and the metaphor of conversation*

As distinct from a real world 'out there', the real world can only ever be our world of experience ... the world in which the individual acts and lives. An individual constructs the world in which he or she lives and we share the meaning of these constructions through communication. My real world is different from your real world and this must always be so.[5] The common ground which is the basis of our ability to communicate with one another, comes about through the use of the common process of perceiving and conceptualising. The process might be common but the end products are never the same. What we share is communication of the worlds we experience, we do not share a common experiential world.

Since it is communication (internal and external) that creates what we call reality, developing a 'shared meaning' (a notion created by the observer)

---

5 / An amusing illustration of how differently we can experience the world because of our acquired habits to do with how we make distinctions and punctuate our world, is the joke told by Paul Watzlawick: '... a man arrives in heaven and finds an old friend sitting there with a luscious young woman on his lap. "Heaven indeed", says the newcomer, "is she your reward?" "No", replies the old man sadly, "I am her punishment!"' (*How Real Is Real?* (1977), p. 62).

is going to involve the participation in the task of all those who will be affected by any outcome. If we accept that living systems are structure-determined systems then communication is a structural coupling of two (or more) individuals in conversation. So to converse is to dance: to turn together in a way that acknowledges the presence of two parties (one of course could and does converse with oneself) and acknowledges the willingness to act together in some mutually acceptable way. The meaning that we are inferring is similar to that found in the original Latin words: *con* ... meaning 'with', and *versare* ... meaning 'to turn'. The actual dance, the experience of the conversation, is a unique creation and we have no certainty whatsoever as to what the outcome might be. It is neither a transfer nor a sharing of information. Useful knowledge, knowledge that will lead to satisfying action, is created by the joint action of both parties and encompasses both scientific and aesthetic judgements.

1.3.4 A 'knowledge and information system' does not lead to action

No one has done more towards achieving a robust development of a conceptual understanding of the research–technology transfer interface than Neils Röling. His major conceptual tool, the generation of an Agricultural Knowledge and Information System (AKIS), is the integrated group of people that encompasses scientists, subject-matter specialists, village-level extension workers, and pastoralists. The members of the group (the system) are together for 'the purpose of working synergically to support decision-making, problem solving and innovation' in any specific domain of agriculture (Röling, 1990, p. 1). He proposes the ideal that all major parties in the system engage in all its major functions: 'the generation, transformation, transmission, storage, retrieval, integration, diffusion and utilization of knowledge and information' (*ibid*). It is not surprising that, on the basis of what is actually happening on-the-ground, he concludes that there is still too much emphasis on 'downstream' functions, suggesting the use of a one-way model. Such is the pervasiveness of the dominant conceptual tradition that Röling admits that 'we have no words for the functions to be performed in shifting indigenous knowledge and farmer influence 'upstream' toward the science end of the science-practice continuum' (*ibid*, p. 36). Nothing is more certain than that we are entering unfamiliar intellectual territory as when we realise that we do not have the words to talk about our experience. And without the language, the ideal can not be transformed into purposeful action.

The experiences in agriculture and rural development have parallels in computing and artificial intelligence (AI) research (the metaphors and actions of scientists from the latter were quickly taken up in other disciplinary domains). As Terry Winograd (1997) points out 'the promises of massively increased productivity through knowledge engineering didn't come true' because 'the mainstream AI effort rested on a view of human intelligence and action that implicitly assumed that all of intelligence could be produced by mechanisms that were inherently like the conscious logical manipulation of facts represented in language'. The detailed arguments which refuted this position appear in the book by Winograd and Flores (1987).

1.4 What is second-order R&D?

The method of doing science espoused by Maturana (1988), which we follow for part of our research (Chapter 6), challenges the way of knowing and acting-in-the-world that: (i) sees an objective reality 'out there' (externally independent of the observer) and (ii) conceives humans as possessing an ability to increasingly know and understand such a reality. While we behave *as if* this way of knowing and acting was a possibility, biologically it is not (see Chapter 3).

Second-order R&D is built on our scientific understanding that human beings determine the world that they experience. The application of science demands that we reflect upon how we operate as perceiving and knowing 'observers' who *bring forth* their experiential worlds through the actual functioning of their nervous systems and the cognitive operation of making distinctions: You have to look in order to see![6]

The characteristics of second-order R&D can be summarised as:

- The doing (the praxis) is grounded in the extending of an *invitation* to, and the willing acceptance by, another to join in making a space for mutually satisfying action.
- The reality that is brought forth includes the researcher, to constitute a duality. It is not subjectivity – subjectivity belongs to objectivity (see Box 1.1).
- All participants share the responsibility associated with every outcome.
- It involves the study of relationships, particularly their nature and quality rather than enties or objects.
- As science, it is grounded in the explanation of what is experienced and, unlike philosophy, is not concerned with adherence to, or the explication of, principles. It has no imperative character.

6 / Here Einstein's famous remark to Heisenberg comes to mind: 'It is the theory that determines what we can observe'.

> **Box 1.1 Duality and Dualism**
>
> It is now widely known that light can be treated as both a wave and a particle depending on the experiment we, as observers (or experimenters) have decided to use to observe its behaviour. This apparent paradox, i.e. wave-like behaviour and particle-like behaviour, was described for many years as the 'wave–particle dualism', which implied they were separate or opposite phenomena. The term used to describe antagonistic or negating opposites is *dualism*, e.g. mind/matter, objective/subjective. Two concepts form a dualism when they belong to the same logical level and are viewed as opposites. The logic behind this dialectic is negation. Reyes (1995) suggests that dualistic thinking is a product of the prevailing objectivist Cartesian world view with its orthodox logic under which we are still brought up. He also suggests that dualisms are responsible for ephemeral and endless debates, e.g. centralisation versus decentralisation. Dualistic or either/or thinking can often represent a trap in our thinking.
>
> It was not until it was recognised that phenomena we observe in 'nature' are not independent of our observing that this paradox was resolved by appreciating that wave-like and particle-like behaviour were complementary behaviours that constitute a duality. Taken as a whole they do not negate each other but constitute a unity or whole. A commonly used example of a duality taken from ecology is the predator–prey relationship. Two concepts form a duality when they belong to two different logical levels and one emerges from the other. The logic behind this dialectic is self-reference. The following pairs are examples of dualities: environment/system; control/autonomy; constraint/freedom; 'what'/'how'). When recognised as complementary pairs the discussion is potentially more rewarding and exciting.
>
> NB. The term 'dialectics' comes from the Greek *dialecktike (techne)*, the dialogical art, which in turn derives from *dialegesthai*, which means talking together, holding a dialogue. Etymologically, dialectics thus means the art of unfolding meaning of a word or idea through a conversation in which two or more persons argue pro and contra. Dialectical thinking is open and dynamic in contrast to formal-logical thinking, which proceeds in a linear and unreflective manner and is thus closed and static.
>
> Source: Open University (1997). *Environmental Decision Making: A Systems Approach*. Adapted from: Reyes, A. (1995). A theoretical framework for the design of a social accounting system. PhD Thesis, University of Humberside, UK.

The need for explicit contextual grounding is at the heart of this conceptual development. This contextual grounding has to do with an increasing understanding of the social construction of the very concepts of the 'research–development relationship' and 'rangeland'. A contextual science is increasingly based on exposing the workings and limits to disciplinary understanding and on exposing a need for an ethic coherent with the capacity to respond to situations.

A problem with 'first-order' science is that it assumes that rangelands can be studied in isolated fragments and that an understanding of the whole can be gained by simply aggregating the detailed understandings. Second-order science (R&D) accepts that real systems (rangeland *per se*) are essentially unknowable and that all science can do is to generate models of reality. And models of 'rangeland' and models of 'research and development' are just that: models.

The essence of this new knowing is that all phenomena are *self-referential* (built mirror-like, by reference to themselves) and *dialectical* (the dynamic relationship between the selected elements brought into experience by the act of making a distinction – see Box 1.1). The notion of self-reference is in direct contrast to our traditional values of 'being objective' or holding a 'neutral position'. The intention behind these values is a very worthy one, it is just that it is scientifically impossible to maintain. Von Foerster (1971) summarises the historical shift as follows:

'Self-reference' in scientific discourse was always thought to be illegitimate, for it was generally believed that The Scientific Method rests on 'objective' statements that are supposedly observer-independent, as if it were impossible to cope scientifically with the referee in the reference, the observer in the description and the axioms in the explanation. This belief is unfounded, as has been shown by John von Neumann, Gotthard Gunther, Lars Lofrgen and many others who addressed themselves to the question as to the degree of complexity a descriptive system must have in order to function like the objects described, and who answered the question successfully

(pp. 239–240).

Much earlier, 1932, the physicist Planck put it his way: 'Science cannot solve the ultimate mystery of nature . . . because, in the last analysis, we ourselves are part of nature, and therefore, part of the mystery we are trying to solve'.

1.4.1 Research as a dialectical relationship

The name for certain sorts of relationships, of which self-reference is an example, is a 'dialectic' (see Box 1.1). Much of our traditional view constructs its knowledge on the basis of dualisms: science as distinct from art; pastoralists as distinct from rangeland; mind as distinct from matter; and so on. The process of a dialectic encourages us to continuously re-connect dismembered dualisms. The new epistemology, often called cybernetic epistemology (Keeney, 1983), or second-order cybernetics (Howe & von Foerster, 1974), uses a dialectic which continuously exposes both sides of our distinctions (e.g. rangeland and pastoralists) and keeps them connected in a recursive way: the rangeland creating the pastoralist and the pastoralist creating the rangeland. In a real sense, the pastoralist is the rangeland and not an actor in it as though the rangeland existed independently of the pastoralist. The practical implications of this epistemology are far reaching. No longer is the pastoralist the 'problem', or the degraded ecosystem the 'problem'. The pastoralist and the rangeland are now seen as a complementary pair: they are distinct but *related*. The dialectical process allows us to look at the quality of the relationship as a 'variable' in research. In second-order science it is not objects that command attention but the relations between them.

What became increasingly clear to the students of 'systems' was their own role as observers. It was the observer, by means of making a distinction, who specified that a system was a unit distinct from its background. It was the observer who then attributed to both system and background their respective properties, properties which justify the act of seeing them as separate. This act of making a distinction is the most basic cognitive act; it is what is at the heart of any investigation of knowledge. So, what is seen to constitute a system is a decision made by an outsider who, for reasons of his or her own, wants to explore a set of relationships.

1.4.2 *How first-order and second-order R&D are related*

Second-order R&D in no way replaces the validity of first-order R&D. Rather, they are related in complementary fashion. In fact, second-order R&D is the context of the first-order. The relationship between the two is itself an example of a duality constituted through a dialectical process: there is 'science', and there are the 'processes leading up to' science. There are pragmatic strategies gleaned from first-order thinking which are contextualized by the systemic wisdom of second-order thinking. With second-order R&D we are moving towards setting a context for change which necessarily complements strategic and consciously pragmatic strategies of intervention (see Umpleby, 1994). It is the more encompassing, building upon the insights and strategies gleaned from the first-order models.

What is being proposed is not an interdisciplinary mingling of the 'two cultures', rather it is a new science. It is a contextual, systemic, and dialectical science.

1.4.3 *Objectivity is replaced by responsibility*

Because of the active focus on the social construction of knowledge, technology, and the very 'doings' of R&D, second-order/contextual science gives attention to people's participation in terms of *power* and *control*. There are the very concrete issues of development for whom? Who benefits, who loses, and who has increasing control of resources and decisions? This is not a simple matter of redefining the problem as this would imply staying with the old framework. What this contextual science looks like on-the-ground would be:

- Evidence of emancipation from powerful authority (including dependency on the disciplinary knowledge of 'science' . . . scientism!).

- Evidence of greater empowerment through collaboration.
- Collaboration based on mutually accepted difference (each person's reality is as valid as another's even though it might not be seen to be as desirable).
- Collaboration is based on shared enthusiasms-for-action.
- Recognition of a complementarity in personal skills or access to resources.

1.5 Precursors of the second-order approach

As academics we both were initially attracted to an experiential and student-centred approach to agricultural education in the early 1980s (see Ison, 1990). This approach stressed the importance of structuring the educational programme around the student's learning needs and the practical problems currently facing the agricultural industry, rather than a prescriptive curriculum based on building blocks of accepted knowledge. The attraction of this educational philosophy and practice resulted from our personal experiences as students with the educational system and from our close working relationships with the rural community. There was a strong sense that learning did not work along the lines espoused by the professional educators. Likewise, there was the belief that knowledge did not flow from the experts to the practitioners.

Coupled with this historical experience were the recent findings from neurophysiological studies shedding new light on the processes of perception and cognition. Then came the complementary models of human communication which presented *meaning* as a relational phenomenon. Meaning is brought about in the interaction and is not present in the head of either the 'sender' or the 'receiver'. This constructivist's view of knowledge formation, information, and learning, began to provide the theoretical underpinnings for a research project involving both authors and aimed at the identification of agronomic problems *in context*, that is, not detached from the more complex social and community issues in which they were embedded. The research developed out of the tradition of rapid rural appraisal (RRA) that had been successfully applied in less developed countries (Khon Kaen, 1987).

This first exploration of RRA in Australia (Ampt and Ison, 1989) took place in central western New South Wales (Forbes shire). One of the key outcomes of this study was the impetus to start work on an alternative model for participatory agricultural research and development in Australia. Following on from this exploratory research, the authors were commissioned by the then Australian Wool Corporation to undertake a critical review of rural extension encompassing both its theory and practice (Russell et al., 1989).

Probably the most significant finding that flowed from the RRA study and the subsequent Critical Review was that all participating farmers and graziers, who were representative of the diverse range in a large agricultural district, were *enthusiastic* about what they liked doing. If they wanted to do something they became well-informed about the relevant issues and did the task competently. They did not have to be educated or persuaded by any outside source when it was their learning need that they were responding to. This is taken up again in Chapter 6.

The elements of our second-order R&D were beginning to fall into place. Next came a collaborative research project by one of the authors, between farmers and the extension services in the Swiss Emmental (Scheuermeier and Ison, 1991). This research evolved from the 'farmer first' tradition in which rapid rural appraisal had frequently been employed using multidisciplinary teams and local people in the 'identification' of problems for research and development. This research, however, moved beyond much of the RRA experience at that time to encompass aspects of the emerging theoretical position described in this chapter. It also attempted to move beyond multidisciplinary to genuine interdisciplinary collaboration by attempting to explore each other's perceptions of what we experienced and how we interpreted these. Through the process of collaboration, and the acceptance of the worldviews of those involved, issues were brought into being, and formulated. In the Swiss research, understandings, derived through the process of semi-structured interviewing, of farmers' histories and present circumstances were used as a basis for them to identify potential actions which might sustain their involvement in farming. Later, in a community setting, farmers were able to join with others in the community who shared common enthusiasms for future action.

Issues identified for future action included: (i) new products; (ii) wood chipping from forest by-products for domestic energy; (iii) machinery and labour sharing; (iv) farm–household diversification; and (v) further information and training. The community forum also provided an opportunity for women to come together and participate for the first time. This resulted in the formation of a women's support group and the public articulation by the group's spokeswoman of the incredible pressure they were under and the need for men to change their ways of working.

These experiences and our emerging conception of second-order R&D led to our involvement with pastoralists, conscious that so often the interventions resulting from first-order R&D have led to the 'administration of carrot and stick incentives . . . (and a failure to) begin to develop systemic frameworks for thinking about things' (Fisher, 1990). As will be illustrated in subsequent chapters, the contextual approach aims at increasing the capacity of pastoralists to respond and offers a clear alternative to the carrot and stick approaches to effecting human action.

1.6 Concluding comments

As we outlined in the introduction to this section, it was possible to recognise three streams of inquiry which we found necessary to pursue at the start of our research. This chapter has referred to all three but has had as one focus a review of the intellectual traditions which have given rise to our very conception of rangelands, rangeland management and rangeland science. A second tradition, which gives rise to the meaning we give to human communication, and from this to information, knowledge and understanding, is also explored. For many readers this may be the most challenging set of ideas because it runs counter to the current common and everyday understandings and to the language that is used. For this reason, and because its relevance is universal, we feel it is important to explain this tradition in more detail in the next chapter.

References

Ampt, P.R. and Ison, R.L. (1989). Rapid rural appraisal for the identification of grassland research problems. *Proceedings XIV International Grassland Congress*, Nice, pp. 1291–92. Association Francaise pour la Production Fourragere, Versailles.

Andrew, M.H. (1988). Rangeland ecology and research in the tropics in relation to intensifying management for livestock industries. *Proceedings Third International Rangelands Congress*, New Delhi, 6–11 November.

Baden, J. and Stroup, J. (1977). Property rights, environment quality and the management of National Forests. In *Managing the Commons*, ed. G. Hardin and J. Baden, pp. 229–40. W.H. Freeman, San Francisco.

Chambers, R., Pacey, A. and Thrupp, L. eds (1989), *Farmer First*. Intermediate Technology Publications, London.

Ellis, J.E. and Swift, D.M. (1988). Stability of African pastoral ecosystems: alternative paradigms and implications for development. *Journal of Range Management*, **41**, 450–9.

Fisher, F.G. (1990). Bicycling out of the greenhouse: a problem of substance. *The Trumpeter*, **7**, 38–44.

Fortmann, L. (1989). Peasant and official views of rangeland use in Botswana. *Land Use Policy*, July, pp. 197–202.

Gilles, J.L. (1985). Slippery grazing rights: using indigenous knowledge for pastoral development. In *Arid Lands: Today and Tomorrow, Proceedings of an International Research & Development Conference*, Tucson, Arizona, USA, Oct. 20–25, pp. 1159–66.

Hopcraft, P.N. (1981). Economic institutions and pastoral resource management: considerations for a development strategy. In *The Future of Pastoral Peoples*, ed. J.G. Galaty *et al.*, pp. 224–43. International Development Research Centre (IDRC), Ottawa.

Howe, R. and von Foerster, H. (1974). Cybernetics at Illinois. *Forum*, **6**, 15–17.

Howells, J. (1990). Economic growth: the globalisation of research and development: a new era of change? *Project Appraisal*, **17**, 273–85.

Hunter, J.P. (1990). SADCC workshop discusses problems and prospects for improved rangeland management. *ILCA Newsletter*, **9**, 4–5.

Ison, R.L. (1990). *Teaching Threatens Sustainable Agriculture.* International Institute for Environment and Development, Gatekeeper Series No. 21. 20 pp.

Jiggins, J. (1993). From technology transfer to resource management. *Proceedings of the XVII International Grassland Congress 1993*, pp. 615–22. New Zealand Grassland Association, Palmerston North.

Keeney, B. (1983). *The Aesthetics of Change.* Guilford Press, New York.

Khon Kaen (1987). *Proceedings of the 1985 International Conference on Rapid Rural Appraisal.* Khon Kaen University, Thailand.

Krippendorff, Klaus (1993). Major metaphors of communication and some constructivist reflections on their use. *Cybernetics & Human Knowing*, **2**, 3–25.

Lamprey, H.F. (1983), Pastoralism yesterday and today: the overgrazing problem. In *Tropical Savannas: Ecosystems of the World*, ed. F. Bourliere, pp. 643–66. Elsevier, Amsterdam.

Lane, C. ed. (1998). *Custodians of the Commons. Pastoral Land Tenure in East & West Africa.* Earthscan, London.

Lane, C. and Moorehead, R. (1994) *Who Should Own the Range? New Thinking on Pastoral Resource Tenure in Drylands Africa.* Pasture Land Tenure Series No. 3, IIED Drylands Programme, IIED, London.

Latour, B. (1987). *Science in Action: How to Follow Scientists and Engineers Through Society.* Open University Press, Milton Keynes.

Leach, M. and Mearns, R. eds (1996). *The Lie of the Land: Challenging Received Wisdom on the African Environment.* Heinemann and Currey, London.

Le Houerou, H. (1989). *The Grazing Land Ecosystems of the African Sahel.* Springer-Verlag, New York.

Little, P.D. (1982). *The Workshop on Development and African Pastoral Livestock Production.* Institute for Applied Anthropology, Binghamton, New York.

Maturana, H.R. (1988). Reality: the search for objectivity or the quest for a compelling argument. *Irish Journal of Psychology*, **9**, 25–82.

Maturana, H.R. and Varela, F.J. (1980). *Autopoiesis and Cognition: The Realization of the Living.* D. Reidel, Boston.

Maturana, H.R. and Varela, F.J. (1988), *The Tree of Knowledge: The Biological Roots of Human Understanding.* Shambhala, Boston.

Ndagala, D.K.A. (1985). Local participation in developing decisions: an introduction. *Nomadic Peoples*, **18**, 3–6.

Open University (1997). Environmental Decision Making: A Systems Approach. MSc Module, The Open University.

Pearson, C.J. and Ison, R.L. (1997). *Agronomy of Grassland Systems*, 2nd Edn. Cambridge University Press, Cambridge.

Planck, M. (1932). *Where is Science Going?* Norton, New York.

Reddy, M.J. (1979). The conduit metaphor – A case of frame conflict in our language about language. In *Metaphor and Thought*, ed. A. Ortony. Cambridge University Press, Cambridge.

Reyes, A. (1995). A theoretical framework for the design of a social accounting system. Unpublished PhD Thesis, University of Humberside, U.K.

Röling, N. (1990). The agricultural research–technology transfer interface: a knowledge systems perspective. In *Making the Link: Agricultural Research and Technology Transfer in Developing Countries.* Westview Press, Boulder.

Russell, D.B., Ison, R.L., Gamble, D.R. and Williams, R.K. (1989). *A Critical Review of Rural Extension Theory and Practice.* University of Western Sydney, Richmond.

Russell, D.B., Ison, R.L., Gamble, D.R. and Williams, R.K. (1991). *Analyse Critique de la Theorie et de la Pratique de Vulgarisation Rurale en Australie.* INRA, France. 79 pp.

Sanford, S. (1983). *Management of Pastoral Development in the Third World.* John

Wiley, New York.
Sanford, S. (1995). Improving the efficiency of opportunism: new directions for pastoral development. In *Living with Uncertainty. New Directions in Pastoral Development in Africa*, ed. I. Scoones, pp. 174–82. Intermediate Technology Publications, London.
Scheuermeier, U. and Ison, R.L. (1991). *Together Get a Grip on the Future: A RRA in the Emmental of Switzerland*. RRA Notes, IIED, London.
Scoones, I. and Thompson, J. (1994). *Beyond Farmer First: Rural People's Knowledge, Agricultural Research and Extension Practice*. Intermediate Technology Publications, London.
Scoones, I. (1995). New directions in pastoral development in Africa. In *Living with Uncertainty. New Directions in Pastoral Development in Africa*. ed. I. Scoones, pp. 1–36. Intermediate Technology Publications, London.
Shannon, C. and Weaver, W. (1949). *The Mathematical Theory of Communication*. University of Illinois Press, Urbana, Ill.
Sless, D. (1986). *In Search of Semiotics*. Croom Helm, London.
Umpleby, S.A. (1994). *The Cybernetics of Conceptual Systems*. Institute of Advanced Studies, Vienna, Austria. 15 pp.
UNESCO (1993). *Statistical Yearbook*. UNESCO, Paris.
von Foerster, H. (1971). Computing in the semantic domain. *Annals of the New York Academy of Science*, **184**, 239–41.
Watzlawick, P. (1976). *How Real Is Real?* Random House, New York.
Winograd, T. and Flores, F. (1987). *Understanding Computers and Cognition: A New Foundation for Design*. Addison Wesley, New York.
Winograd, T. (1997). From computing machinery to interaction design. In *Beyond Calculation: The Next Fifty Years of Computing*, ed. P. Denning and R. Metcalf, pp. 149–62. Springer-Verlag.

2 The human quest for understanding and agreement

Lloyd Fell and David B. Russell

2.1 Introduction

There was a Professor in Australia many years ago, with whom one of us was associated, who was remarkably successful at talking to farmers. He always drew a big audience at field days and his discussion groups were packed with people keen to contribute. The local advisory officers were not so pleased with him. Some thought that the farmers 'could not have understood a word he said', so technical was his vocabulary and so imaginative were some of his ideas. It was true he used a lot of scientific terms, but there was also a certain warmth and genuine enthusiasm about him, particularly when engaged in conversation. There were visible signs in the farmers of rapt attention to what he was saying. They seemed to 'follow' him, unlikely though that may have seemed at times. Because they said they understood him so well – and felt that he understood them – the farmers formed a special Foundation which funded his entire research programme for the decade or so he spent in Australia.

The R&D professional, traditionally scientifically trained, seeks to achieve a level of mutual agreement with clients. In the past this agreement was a very conscious pursuit. Today it has a more subtle quality. No longer do we hear: 'I know what is good for you.' The Professor was an enthusiastic proponent of his own experience, but apparently without being committed to expect a precise literal interpretation of what he was saying.

What follows in this chapter is the unfolding of an explanation, essentially based in the biology of cognition, which addresses both the Professor's success as a communicator and the concepts underpinning the collaborative research that is the subject of this book. The thread running through it is the thinking and the experience of the research group. The detail is a conceptual framework which supports this approach. The form of R&D which arises from this combination of theory and practice is very different from the traditional form.

How is it that we can say we understand someone or understand something about the world in which we live? Is there a state of mind or body that we can usefully refer to as *understanding,* or another state that we might call *agreement* or, as a group, consensus? We can make these distinctions because we recognise them directly from our experience, but it is worth remembering at the same time that we are speaking as observers of these phenomena of human interaction. They are both individual and collective experiences and observations. If we want to describe the process which leads to these states, we put the question this way: what is it that we would need to have observed, in others or in ourselves, for us to say that understanding (or agreement) had occurred?

In the traditional approach to thinking, learning or researching, the circularity inherent in this question – in trying to understand understanding – tends to overwhelm us. There is a new approach, however, that comes from a particular branch of biology known as second-order cybernetics or the 'biology of observing systems' (von Foerster, 1984), which has arisen by breaking out of the mainstream biological tradition. This new branch of biology deals particularly with perception and cognition and forms part of the emerging tradition that is referred to as second-, rather than first-order, in Chapter 1.

The ability to reach agreement is important to us in many different ways and consensus is often esteemed, especially in political circles, but there seem to be different kinds of agreement and some of these may not necessarily involve an understanding. There are also two distinct aspects of the notion of understanding that we wish to consider here. One is understanding how something works, that is, the logical causal relationships which make it happen; the other is understanding one another. Our culture is so intensely immersed in the first-order tradition that more attention is given to the first aspect. Understanding is generally thought of as entailing processes of reason or rationality and concerned with the technology and the information at hand. But this attitude underestimates and devalues the crucial importance of human relationships themselves and the underlying influence of human emotions and feelings as determinants of what actually happens in our working together.

The thread guiding the research project described in Chapter 6, began with the idea that what really matters is that we become an active partner in forming a mutually beneficial relationship with our clients.

The new body of biological theory has led us to the view (see Fell and Russell, 1994) that a satisfying experience of understanding does not result from the transfer of information, nor from the force of reason arising from a compelling argument (the truth), but it does result from some other properties of our biological interaction. Furthermore a satisfying experience of agreement is closely linked to this human process of understanding. What these biological properties are and what brought us to this view are discussed in this chapter.

2.2 Second-order cybernetics in relation to communication theory

The business of human interaction is often described as *communication*, the major currency of which is *information*. An examination of the history of communication theory reveals that it has changed profoundly over the last two decades. Two factors influencing this change have been the advent of powerful computers (to better model complexity) and the thinking which is known as second-order cybernetics. The idea of communication as the transmission of unambiguous signals which are codes for information has been found wanting in many respects. Heinz von Foerster, reflecting on the reports he edited for the Macy Conferences that were so influential in developing communication theory in the 1950s, said it was an unfortunate linguistic error to use the word 'information' instead of 'signal' because the misleading idea of 'information transfer' has held up progress in this field (Capra, 1996). In the latest theories the biological basis of the language we use has become a central theme.

Cybernetics, although often applied to the control of machines, has long been one of the foundations of thought about human communication, its central notion being circularity. Cybernetics 'arises when effectors, say a motor, an engine, our muscles, etc., are connected to a sensory organ which, in turn, acts with its signals upon the effectors. It is this circular organisation which sets cybernetic systems apart from others that are not so organised' (von Foerster, 1992). In first-order cybernetics it was the idea of feedback control which mainly occupied the practitioners, but in time the question 'what controls the controller?' returned to view (Glanville, 1995a,b) and the property of circularity became the focus of attention once again.

Second-order cybernetics is a theory of the observer rather than what is being observed. Heinz von Foerster's phrase, 'the cybernetics of cybernetics' was apparently first used by him in the early 1960s as the title of Margaret Mead's opening speech at the first meeting of the American Cybernetics Society when she had not provided written notes for the Proceedings (van der Vijver, 1997). This is a philosophical jump of such proportions that many writers on human communication still choose not to acknowledge it too openly. We do so because we think it provides a bridge from the rather infertile land of communication theory based solely on the idea of information transfer to another still largely uncharted territory where more basic biological mechanisms need to be considered. It requires a loosening of our grip on the supposedly certain knowledge that is acquired objectively, about a reality existing independently of us, and a willingness to consider the constructivist idea (see Mahoney, 1988) that we each construct our own version of reality in the course of our living together.

The virtue of objectivity was that the properties of the observer should be separate from the description of what is being observed. This led to what von Foerster (1992) called the Pontius Pilate attitude of abrogating responsibility because the observer is an innocent bystander who can claim he or she had no choice. The alternative attitude, which seems to be less popular today, is to own a personal preference for one among various alternatives. This is the contrast between first and second-order traditions which was raised in the introduction to this section: am I apart, looking through peepholes, or am I a co-creator of the kind of world in which I live?

It was the biological theory of *autopoiesis* presented by Maturana and Varela (1980, 1987) that enabled the limitations of earlier models of communication and the opportunity for a new approach to be revealed. For years Maturana had wrestled with two questions which he eventually found had a common answer: *what is perception?* and *what is the characteristic systemic organisation of living things?* In this chapter we will describe some of the details of his finding that the circular 'self-producing' property of living organisms was also the 'self-referring' process of cognition whereby the organism maintains itself in its world. Since the advent of the concept of autopoiesis it has been possible to develop explanations of human communication which adequately account for the properties of the observer as well as the observed.

In this way the biology of our use of language has become a central theme. A particular aspect that we will elaborate on later is the use of metaphorical, as distinct from literal, language. Bateson, one of the pioneers of cybernetic thinking, regarded metaphorical language as especially apt for biology because it captured essential similarities of form and pattern (e.g. Bateson, 1991). Another pioneer, Krippendorf (1993), outlined the major metaphors of human communication and a theory of metaphor which described their ability to create new realities by the way they 'organise their users' perceptions'. He said it is the wide variety of metaphors for human communication (e.g. container, conduit, control, transmission, war and dance) that tells us most about its nature; that we can work in many different ways. We have a certain 'cognitive autonomy' in that our understanding is our own affair, but this understanding can be 'creatively reconstructed' whenever the way we are using our language is 'viable'. This viability in our use of language depends on a congruence in the particular metaphors we use. Krippendorf's analysis leads him to the very point that is central in second-order cybernetics: that our understanding and our actions form an inseparable circular unity; we only do as we understand.

2.3 The interplay of language and emotions

As observers we want to consider not only the behaviour or actions of people together, but also their internal body function which enables them to participate in this interaction. Theories of human communication have often conflated these two quite separate domains of explanation into one by trying to reduce units of behaviour to units of physiology or vice versa, e.g. attempts to find biochemical traces of memory in the brain. One important contribution of second-order cybernetics, and Maturana in particular, was to point out that physiology and behaviour cannot be explained in the same terms, but the process by which they are linked is the crux of the whole process whereby humans interact, agree and understand.

In describing behaviour we catalogue our actions and the language we use. In studying body function we have a particular interest in the operation of our nervous system, often at many different levels, and the pervasive underlying current of our emotions. It is the interplay between our use of language and our emotions, viewed in the terms of second-order cybernetics, which provides the basis for an explanation of understanding.

Why is there a need for understanding? What drives this human quest? The need we feel for understanding and agreement arises from our concerns as they vary from time to time. These can range from an obvious fear or desire for change to a more subtle yearning or interest in something that we perceive to be happening to us and around us. These are the *issues* that we discuss, the matters about which we seek some agreement or understanding and, even more so, to feel that we are understood. Debate over issues such as animal welfare or land degradation due to agriculture seems to rise and fall like the tide. Other issues such as the bank overdraft, rising costs and falling incomes, lap constantly at the edges of our human concern. One noticeable thing about the Professor was that he was always ready to converse on the most topical issues of the day.

We invite you to consider: how do issues arise, where do they come from, and what has to occur before they go away again? Focusing on the biology of language, from a position of engagement with our world, we think that all issues arise in our language and they are only ever resolved in our language. Someone draws our attention to a particularly bad case of soil erosion and that is the principal topic of conversation for a time until, for many possible reasons, that topic of conversation will wane once again. This means that we humans can take responsibility for creating all of our issues – not a single one was thrust upon us by an outside agency – and it follows that we have great potential for resolving even the most difficult of them.

An *explanation* of a process such as understanding also takes on a different light when considered in this way. We are continually trying to explain our experiences, at the same time, knowing that our explanation is not the experience itself. We have a propensity to invent discrete entities such as 'mind' and 'body' as characters in the drama we are trying to explain which tends to divert our attention away from seeing it as a relational process, a phenomenon emerging from some dynamics of interaction. It is probably easier to think of explanations as being about something else rather than being themselves processes of human interaction.

An explanation consists of the telling, and accepting, of the story of how this experience happened – of the events or processes which, if they occurred in this way, would result in that particular experience. It is the acceptance which is crucial; without acceptance, the explanation does not exist. Different fields of scholarship use different criteria for acceptance of an explanation and, as our individual ways of thinking vary, so do our criteria for accepting explanations. An explanation can only be valid, therefore, in its particular set of human relations. So we can say that truth, like beauty, exists in the beholding. It follows that understanding is experienced in the genuine acceptance of an explanation. An explanation which is totally accepted is like the pacifier which stops a baby crying – it is an unmistakable sense of satisfaction. Explaining is our major tranquilliser in the western world today and we get our 'fix' through understanding and being understood.

In the R&D research project we believed that we had nothing in particular to tell these clients; rather we sought a mutually satisfying conversation about our respective experiences.

In developing his explanation of *what it is to be human*, Maturana said that we do not just use language, we are immersed in it. Our ever-changing present reality consists of how we describe our experiences to ourselves and one another and we are always explaining and reporting our experience. Furthermore, we act according to our current view of the world. This is quite different from regarding language as a means of communicating or transmitting information using symbols or representations of an independent reality. We are saying that in our use of language we construct our own reality and we humans have evolved our particular manner of living largely through reliance on the use of language as our principal relational dynamic (Maturana and Verden-Zöller, 1993).

2.4 Autopoiesis and engagement

It was around 1970 that Maturana started to see the living system as a particular type of *closed system* – a closed network of molecular production, but producing itself – for which he coined the term, an *autopoietic* system. He took pains to distinguish its *structure*, i.e. the component parts and their molecular relations, from its *organisation*, which is the particular emergent property of the living system as a whole that must be maintained for the system to go on living. Structure may be likened to the individual notes in a musical composition, whereas the music itself arises in the particular pattern of organisation that these notes reveal when a passage is played. In an autopoietic system it is the continually changing structure that maintains the system organisation and conserves its identity or its relationship, as a whole being, to the medium in which it lives.

This is a paradoxical kind of self-regulating system which is closed with regard to its own operation, but of course is open in its connection to its world. It invokes the idea of complementarity (from quantum physics), that the organism exhibits this autonomy in its operation, yet is dependent on its coupling to the environment. A living thing could never be entirely separate from its environment, nor entirely belong to its environment. In our biological explanation we employ Maturana's 'double look' – distinguishing the organism as an entity which is operationally self-contained in order to see more clearly the nature of its connection with the world in which it lives.

It follows that the nervous system is also closed in its operation, which invites us to see the process that we call *cognition* in a very different light. Since about 1950 the prevailing view in cognitive science has been that the nervous system picks up information from the environment and processes it so as to provide a representation of the outside world in our brain. We can now say instead, to paraphrase Varela (1979), that the nervous system is closed, without inputs or outputs – that its cognitive operation reflects only its own organisation – and that, because of this, we are imposing our *constructed* information (or our *meaning*) onto the environment, rather than the other way around.

This implies that our interactions can never be instructive, i.e. unambiguous external signals. They consist of non-specific *triggers* which disturb the system, but do not determine the nature of the response. The operation resulting from the trigger always depends on the internal coherence or arrangement of the respondent at that time. So the nervous system does not operate with representations of the environment – even though it may often appear, to an observer, to be doing so. The simple logic (which we find satisfying) is that all body systems operate strictly according to their own structural dynamics, i.e. according to the operational necessities of their stream of structural change. In Maturana's terminology, they are *structure-determined*.

Both the organism and the medium in which it lives have their own structural dynamics and their own emergent organisational property which is realised through structural change – so it is their *engagement* with one another at any particular time which is the crucial determinant of the life history of the organism. What supports the organism in its world is the coupling of its structure to the circumstances of that world. It is the same between two individuals. As long as both choose to be mutually engaged, the flow of change in one individual will be linked to the flow of change in the other. This can be described as a physical process involving electromagnetic and mechanical forces for sight, sound, smell and touch and it could include more subtle or speculative forms of energy as well. It is to the mechanism of engagement that we must look to explain the dynamics of change occurring during human communication.

What this means for the R&D research project is that we cannot expect to get inside the other's head with new information, but we do aim to engage with them in some meaningful way. The effectiveness of our communication will be dependent on the manner of our engagement.

The mechanism of this engagement is an aspect of human physiology which will be examined below.

2.5 Physiological coherence

All the operations we could ever hope to observe in physiology are potentially involved in this engagement, but at present we cannot explain this fully. The major difficulty is not our incomplete knowledge of the details – it is our inability to express the operational coherence of the total physiological system. For this reason our usual approach in science is to manipulate various components rather than to understand its operation as a whole. This can be spectacularly successful, but often it is not. A promising new avenue in this regard is the network approach to visualising operations of the nervous and immune systems as they operate together (Varela and Couthino, 1991; Booth and Ashbridge, 1993).

The workings of our physiological systems appear to the observer as a cloud of correlations. We distinguish the components as biochemical or molecular entities and we measure their amounts, e.g. hormone concentrations, etc., but it is according to their pattern of relations that we form our explanations. We determine a certain coherence (i.e. a moving together), which we then interpret in different ways depending on our perspective. We mentioned earlier that this takes place in our language and its acceptance as an explanation is relationship-specific – confined to a particular conversation.

Here we speak of coherence as a pattern of relations, not simply a cause–effect sequence such as the cascade of events by which our physiological mechanisms are most commonly portrayed. Recent explanations in physics – a different explanatory domain – distinguish coherence (in condensed and living matter) from incoherence in the following way: 'In the dominant paradigm . . . particles are localised, separable and countable, "know" each other through collisions and external forces and require an external agent to become ordered. In the "coherent regime", particles lose their individual identity, cannot be separated, move together as if performing a choral ballet and are kept in phase by an electromagnetic field which arises from the same ballet' (Del Giudice, 1993). The closer the correlation the stronger the connectivity between any two biochemical entities which we choose to distinguish. We could represent this in terms of the volume of traffic on different parts of a complex highway network – the traffic flow indicating the strength of relations between those particular centres, i.e. the connectivity of the system.

The particular type of physiological correlation which is crucial to engagement is the sensory–effector correlation. The body surface, which is the interface between organism and environment, can be said to have a dual participation in the outer as well as the inner world. In physiology, surfaces are distinguished as *sensory* or *effector* according to their function, i.e. whether they detect external stimuli or implement some action. This is arbitrary, like all distinctions. The 'double look' shows that sensing and effecting are one operation in an organisational sense. The simplest explanation of the process which we call cognition is a sequence of sensory-effector correlations at the organism's surface.

In terms of the R&D research project it appears that the quality of the engagement with clients will be dependent on the respective past histories of similar engagements.

The autonomous operation of the nervous system – the changing relations of activity according to its own structural dynamics – at any moment in time, is capable of a certain configuration of sensory-effector correlations at its surface. The organism's behaviour – its relations with the medium – also creates potential sensory-effector correlations at the interface. Where these two sets of possibilities meet, there is an engagement, or structural coupling, at that moment. The flow continues according to its history of recursive interaction. Each coupling triggers the change which brings about the next possibilities, so the flow of behaviour and the flow of physiology are mutually modulating. The dynamic matching of internal and external sensory–effector correlations constitutes the course or history of our structural coupling.

The flow of structural coupling can be visualised as a tightrope walker maintaining her balance by means of the exquisite structural dynamics of her bodywork intertwining with the precise behavioural dynamics of her

Figure 2.1
The track laid down in living which arises from our theories of the world. (Used with permission from Michael Leunig, 1990.)

footwork on the rope. She and the rope change together as long as their coupling lasts. There are times when the relationship is shaky and times when it is slick and smooth. Similarly the path – or railroad track – which we lay down in living is sometimes narrow and uneven, sometimes broad and straight. The Australian cartoonist and poet Michael Leunig captures, in one of his cartoons, the meaning we have when we use this metaphor (Figure 2.1).

2.6 Structural coupling and the quality of life

The difference between a smooth or bumpy ride through life is a manifestation of this quality of structural coupling. Although autopoiesis – maintaining organisation – is an all-or-none phenomenon, the degree or quality of engagement seems to vary. While it is true that an organism must fit with its world to go on living – always conserving its adaptation – its grip on life, or its match with the world, appears to us to wax and wane. Living is achieved somewhere between a perfect match and no match at all – either of which would constitute a loss of biological identity. The issue of biological fitness is a relational dynamic which can be represented in terms of structural coupling.

Cyberneticians have pointed out (von Glasersfeld, 1985) that the complementary aspect of autonomy is the necessity, in interaction, to make do with whatever is at hand. It is not necessarily the best fit which occurs in the course of structural coupling – it is whatever connection will work for the time being, i.e. whatever will enable the adaptation to be preserved in that moment. This makes one appreciate that the course of one's life is subject to many vagaries in the delicately balanced connection between our physiology and behaviour.

It is also salutary to note that the idea of choice, like understanding or awareness, arises in the reflection that we make about our experience. It is a commentary on what has happened rather than the happening itself. Our ability for reflection is extraordinarily powerful because, once we have reflected (in our language), we are cognitively different – our physiological coherence has changed – and our opportunities for structural coupling (and therefore the direction of our lives) have changed.

The quality of structural coupling can also be observed in studies of animal behaviour. In research with farm animals (Fell and Shutt, 1989; Fell, 1992) the idea has been developed of behavioural *confidence* as a measure of the quality of structural coupling over a period of time. If the history of structural coupling is one of diminishing opportunities in the environment and declining physiological coherence – whichever 'leads' they will both occur – this will be reflected in a diminished behavioural repertoire for the animal. Certain behaviours that would have been expected to occur in that situation have disappeared. This loss reflects a deterioration in the structural coupling.

Intensively housed farm animals provide examples of this. What is often called 'learned helplessness' is a chronic loss of confidence. What is commonly known as *stress* can also be expressed as a loss of confidence, i.e. having fewer behavioural options, which corresponds with Bateson's definition of stress: 'a lack of entropy ... the organism lacks and needs flexibility, having used up its available uncommitted alternatives' (Bateson, 1980). Animals which lack confidence, or have suffered a reduction in confidence through bad handling, stress or some malaise, are cognitively disadvantaged and this has implications for their subsequent behaviour and also for their health (Gates *et al.*, 1992).

The term *confidence* can also be applied more generally to quality of life. We speak of confident behaviour when there seem to be many options available – few barriers or restrictions to behaviour – an openness to accept the life stream as it is. When this happens our behaviour is often said to be related in some way to our emotional state. Being 'in a good mood' is the natural accompaniment to confident behaviour.

In the R&D project we are interested in what emotions we and our clients are experiencing because we know intuitively that this has some bearing on whether our interaction will be a mutually beneficial and satisfying experience. Our experience tells us that an air of behavioural confidence based on a good quality of physiological engagement is an important attribute which could greatly facilitate our communication.

The way in which emotions and the use of language are intertwined is a crucial aspect of the biology of cognition as it applies to human understanding and agreement.

2.7 Emotions

Maturana (1988) coined the term *emotioning* to distinguish different physiological dynamics by means of observations in the domain of behaviour. He said that emotioning was a *bodily predisposition to action* and that certain characteristics of behaviour could be used to distinguish certain emotions. Maturana said that love is the easiest emotion to characterise in humans because it is seen in the most trusting and respectful kinds of behaviour. In contrast, fear is an emotion which sets an aggressive style of behaviour and also constrains the spectrum of possible behaviours. Similarly, a fearful animal will show a greatly reduced behavioural repertoire, i.e. has lost much of its *confidence*. An analogy used by Maturana is that of a motor car whose structure is in reverse gear so that it does not have forward motion in its realm of possible actions.

This assessment of emotioning (the flow of emotions) is indirect and qualitative, but extremely useful, nevertheless. It is in our describing of emotions that metaphorical language becomes so important as we rely heavily on previous experience and intuition. Charles Darwin was one of the keenest observers of detail in all of biology, yet when he came to make explicit descriptions of what he called the 'emotions of animals' he relied upon drawings, particularly of their facial expressions (Darwin, 1965: 1872). Bateson (1991) referred to the qualitative as pattern rather than number. He regarded patterns of organisation and relational symmetry as characteristically biological and to do with the mind (Bale, 1995).

Kövecses (1990) addressed the question: how do people understand their emotions? He said that 'emotion concepts' have a distinctive metaphorical structure in our language and that these metaphors of emotion 'yield such an unambiguous understanding that they can be seen to represent a coherent cognitive model.' To speak about emotions in the abstract actually sacrifices precision – psychological theories about emotion have little consistency among them – but careful observation of our language shows that metaphor is the vehicle by which we reach agreement about such basic states and patterns as emotions.

Metaphors of communication, discussed in detail by Krippendorff (1993), are seen to be 'vastly more powerful' when we wish to use language, not to represent an external reality, but to organise our experience and interact with one another. They are not merely 'embellishments in language, they affect the users' perceptions and actions'. They convey structural similarities and also have 'entailments' (Lakoff and Johnson, 1980) which organise far beyond the initial similarity and, in Kövecses' view, help to organise the emotion concept itself. Therefore the 'shape' of our understanding is an emotional pattern that is determined by the metaphorical structure of our language.

In the R&D research we needed to become aware of the extraordinarily powerful self-organising properties of the metaphorical language in which we were engaged together with our clients.

The intertwining of emotional experience with words and actions creates something that we recognise as important even though we rarely try to explain it.

2.8 Understanding realised and observed

Watching a child's face when his mother reveals a lost toy – sharing the 'joy of movement' glimpsed in Van Gogh's 'Starry Night' or heard in Beethoven's first piano concerto – are memorable experiences. A sigh between two lovers in their parked car, as the radio pulses shared triggers of contemporary music, is another form of mutual knowing. The aha! experience in speaking with a counsellor, watching a film, or solving a mathematical puzzle; all have a quality of emotional satisfaction about them which we call understanding. Even more pragmatic examples such as finding the directions you need on a map or following a recipe to make a cherry pie are undeniably emotional experiences.

These experiences consist of an intense behavioural interaction which, even without words, occurs as language. However, it is not necessary for any information to have been transferred from one person to the other – as the common explanation implies. Culturally, this is called the 'information age' and we are supposed to come together to exchange information rather than to interact. Thus we value our rationality far above our emotionality since one can process bits of information whereas the other cannot.

This rather mechanical picture does not fit comfortably with the latest biological theories about communication nor with the idea of understanding as a fundamental human quest. Maturana and Varela have explained cognition, not as an information-processing operation, but as a constitutive mechanism of all living things. In knowing ourselves we come to know our world. This explanation is derived directly from experience, rather than from the artificial domain of computer processing. Thus it accommodates both our emotions and our use of language.

From the biology it is apparent (Mingers, 1991) that language is essentially connotative rather than denotative. Especially in science, we generally act as if the words denoted an external reality which existed independently of us. It is convenient and often profitable to do so to explain simple causal relationships. But this expedient turning-a-blind-eye to the connotative nature of our language also obscures the explanation of more fundamental experiences such as understanding.

The fact that we often appear to be more or less in agreement about the meaning of a word or scientific concept is a testament to our ability to work together, not a proof that such an entity exists in reality. We have to say that the meaning of something is not in the words – nor in what they describe – it exists in us, as we relate to that something. So it is context-dependent, meaning different things at different times, even for the same person.

From this it follows that meaning is not transferable – it arises in the course of conversation. The pioneering physicist, David Bohm, compared what he called *dialogue* with a discussion (Bohm, 1990). In a discussion the idea of analysis is paramount and we may not progress 'far beyond our various points of view.' On the other hand, a dialogue is 'a stream of meaning flowing among and through us and between us. This will make possible a flow of meaning in the whole group out of which will emerge some new understanding... this shared meaning is the "glue" or "cement" that holds people and societies together'.

Therefore an experience such as understanding cannot be validated independently of us as observers. We can reach a form of agreement as observers about what we will choose to call understanding, but we are speaking of a knowledge which arises in our conversation – in our living together – not directly through the properties of something independent of us. It is in this way that our whole culture arises through networks of conversation leading to widespread agreement about concepts and values and a comfortable ability to live together with a certain amount of mutual understanding.

Because the flow of our language and our emotions are so delicately interwoven, it follows that, without emotional matching, a semantic congruence could not occur. The meaning which is formed will only match when the emotion matches. Only when we dance in the flow of one another's emotions can we experience understanding. Then we are moving in the same stream – cognitively flowing together. The roots of 'conversation', *con versare*, mean 'turn together' – suggesting dancing.

What we need to have observed in order to agree that understanding had occurred is a harmony of emotions (often metaphorical) underpinning a particular use of language which is fundamentally satisfying. The work of the Professor was misjudged by his critics, who thought that understanding could not have occurred because their idea of the technical meaning of the words and concepts was not shared by the farmers. But the farmers had an emotional rapport with the Professor which engendered trust and the confidence to recognise options and possible solutions to their problems.

To a certain extent this happens all the time. Everything we do in our relationships at work and play, in our families, clubs and institutions, manifests itself in a certain level of understanding of one another. We could not co-exist successfully without this kind of understanding.

We also understand many things about the world in which we live. The process is essentially the same even though it is experienced on one's own. To understand is to grasp a configuration of relationships that exists in our world as we see it. An emotional comfort arising from a particular use of language typifies the experience. This is usually associated with an increased ability to manipulate situations and make things happen, but that is not all there is to it. Very often we also need to be able to share this with another person. To know that someone else sees it in a very similar way is affirming and empowering.

There is a subtle, but very important, qualitative difference between understanding linear causal relationships that explain a supposedly simple mechanism (how it works) and the more systemic form of understanding of a network of coherences that occurs in normal human interaction.

2.9 Understanding, agreement, enthusiasm and confidence

In human co-existence, the quality of an action is normally recognised in terms of its emotional context, e.g. a churlish, but obedient child who 'sits up straight' does so quite differently from an excited, eager child awaiting some reward. This is an evaluation of the person's emotions which we carry out intuitively in the course of our understanding. It also illustrates the fundamental difference between understanding and agreement. Whereas an understanding between two people cannot occur without an emotional accord, agreement certainly can and often does. What we call an agreement is a particular action in language that serves to coordinate our actions, but need not have an emotional basis for even the slightest understanding between the parties involved.

Typically, at a meeting, we agree to go about some joint task in a certain way and this arrangement is sufficient to get the task completed. Different jobs are assigned to different people to avoid duplication of effort and diverse activities are coordinated. There is an obvious practical utility in being able to reach agreement or consensus without the need for understanding, but there is also a cost. Such an agreement can be at the expense of personal and social wellbeing and can also be counterproductive when the subsequent actions are either half-hearted or even destructive.

This book has arisen from a concern that what is known as technology transfer in agriculture is less effective than it could be. It is obvious that the metaphor of direct information transfer does not match the human experience. In the business of R&D it appears that the more determined we are to achieve an information transfer, the less likely it is that effective communication will be achieved.

In contrast, Fell and Russell (1991) and Russell and Ison (1993) discussed the idea of a 'second-order research and development' based on the conversation among the clients, their families and the researchers and the idea of commitment. Winograd and Flores (1987) spoke eloquently of understanding as pattern recognition, conversation as the vehicle for genuine commitment, and 'enthusiasm for action' as the crucial element in corporate success. The emotion of *enthusiasm* was recognised in the early stages of the research described in this book (Webber *et al.*, 1992, Ison, 1993, Russell and Ison, 1993) as being a crucial element in successful technology transfer. The idea of *enthusiasm-for-action* as a central theme is discussed in Chapter 6.

Our R&D experience was that, if we were tuned in to the dynamic interplay of language and emotion in a mutually satisfying engagement, then successful communication was close at hand.

Maturana has pointed out that we are rational/emotive animals, but it is not our rationality which distinguishes us from other animals – it is the way our rationality and emotions braid together. We are animals that naturally use reason to justify our emotions. The prevailing culture today, especially in science, is to deny and denigrate emotionality, particularly when stacked against the supposed ideal of a precise and accurate understanding. This has many consequences in the relationships which make up daily life.

When we start to discuss our emotions with one another a further difficulty arises. What we describe as our feelings does not necessarily correspond with our *emotioning* because it is a reflection that we make – a commentary about our experience – which is shaped in our language like any other explanation. A wife may laugh to hear her husband saying 'I feel fine' in a loud and angry voice – a girl may say 'it meant nothing' with tears in her eyes – many incongruities may arise in our feelings.

The reputed unreliability of feelings, combined with a craving for immediate technological remedies (actually to treat uncomfortable feelings!), have contributed to a profound devaluation of the knowledge which is our experience of living. Yet the extraordinary clarity and precision by which poetic images enable us to communicate shows that it is through our experience of living – not our theories – that we can know the satisfaction which we call genuine understanding. Because our interaction is a mutually triggering experience, not an information transfer, the songs we sing together are as important to our understanding as the discourse which we have. Pictures can connect us powerfully, providing personal meaning via sharp, shared, triggers in our rational/emotive dance. What would become of human understanding if it were not for the theatre, art and music which we create and enjoy together?

This has implications in a number of domains. It seems obvious that an educator's work will be less successful when the student is not emotionally inclined towards the educational task. The feeling of wanting to listen is probably the main prerequisite for any form of education. The educator who tries to conceal his own emotions so as to give the most rational explanation of his topic may be sacrificing opportunities for understanding to occur. The ability of a schoolteacher to establish an emotional rapport, or mutual respect, with her pupils may explain much of the variation in classroom performance.

Good salespeople everywhere know about matching emotional shapes. In real estate, the salesman need not even be entirely factual about the details of the house, but must know with whom it will fit – who will have an 'understanding' with the house and want to buy it. Experience at observing human behaviour – noticing the emotional reactions of a client during the inspection of a house – is invaluable.

In this book we are dealing with the application of science to agricultural issues. The way that Maturana speaks of science, not as a means of discovering an objective reality which is independent of us, but as a useful way of operating in the world, enables us to acknowledge that scientific data is valuable because it helps to shape the meaning which we form in the course of our conversation, but it does not determine that meaning. Thus scientists need not be – nor do they function as – final arbiters on any community issue, but their contribution is important, nevertheless.

This sharpens the focus on ethics and a sense of personal responsibility in the business of doing science. The wry quote attributed to von Foerster (Glasersfeld 1985) sums it up: '. . . invoking objectivity is abrogating responsibility; hence its popularity!' We surely have great opportunities in science if we are not simply talking 'facts', but offering scientific interpretation –

claiming only to be custodians of valued scientific data which can make a helpful contribution to the networks of conversation that make up our human culture and, ultimately, to our continued existence.

As technically trained, scientific, R&D communicators today we found we could have confidence in our ability to provide a scientific interpretation, but that genuine understanding and agreement between the parties, based on emotional synergy achieved through language, was the key to truly collaborative action.

The confidence that comes with understanding opens up more choices and creates more possibilities in human interaction. It appears that a confident person can listen more closely and notice more details of the interaction than one who is lacking in confidence. It seems easier to understand a confident person – to move together coherently – just as it easier to dance with a natural, confident dancer. In public speaking, we may notice that the way we speak is every bit as important as the content. This is simply acknowledging the value of the interaction itself. Interacting leads to more interacting; there is a flow.

It is likely that the confidence which people show as human beings is being sapped by too much reliance on the rational element in understanding and a denial of the emotional basis which pervades all human interaction. A lack of confidence restricts our options and limits the potential of our human co-existence. The most memorable thing about listening to the successful Australian Professor was the liberating feeling that there were possibilities opening up, more opportunities for bringing about change, other ways of looking at this complicated world in which we all want to live comfortably and confidently.

If an explanation 'works', i.e. some understanding arises in the biological interaction, it brings some satisfaction and also an appetite for more of the same. This process contributes to the building of our confidence which, in turn, points towards that ethical imperative which von Foerster offered to us (e.g. von Foerster, 1992): 'act so as to increase the number of choices'.

In the very human business of agricultural extension and rural development there are many situations in which a lack of confidence and discrediting of the emotional aspect of the experience could lead to a limited understanding by all parties and rather ineffective agreements about future courses of action. In such cases everyone is acting in a guarded fashion to limit the number of possibilities available, being fearful of the uncertainty, lacking in trust. However, if the underlying biological principles are observed in the course of this work, opportunities also abound for the creation of genuine understanding, committed agreements and *enthusiasm for ac-*

tion so that there will be a sense of empowerment in the people concerned and enhanced confidence to go on increasing the number of choices available and enriching the human experience.

References

Bale, L.S. (1995). Gregory Bateson, cybernetics and the social/behavioural sciences. *Cybernetics & Human Knowing*, 3(1), 27–45.
Bateson, G. (1980). *Mind and Nature: A Necessary Unity*. Bantam Books, New York.
Bateson, G. (1991). *A Sacred Unity: Further Steps to an Ecology of Mind*, R.E. Donaldson. HarperCollins, New York.
Bohm, D. (1990). *On Dialogue*. Transcript of meetings and seminars edited by David Bohm. David Bohm Seminars, Ojal, CA. 41 pp.
Booth, R. and Ashbridge, K. (1993). A fresh look at the relationship between the psyche and immune system: teleological coherence and harmony of purpose. *Advances*, 9, 4–65.
Capra, F. (1996). *The Web of Life. A New Synthesis of Mind and Matter*. HarperCollins Publishers, London.
Darwin, C. (1965). *The Expression of Emotions in Man and Animals*. University of Chicago Press, Chicago. (Originally published 1872.)
Del Giudice, E. (1993). Coherence in condensed and living matter. *Frontier Perspectives*, 3, 16–20.
Fell, L. and Shutt, D. (1989). Behavioural and hormonal responses to acute surgical stress in sheep. *Applied Animal Behaviour Science*, 22, 283–94.
Fell, L. (1992). Does it matter what sheep think? In *Rural Science Annual 1992*, pp. 45–7. University of New England, Armidale.
Fell, L. and Russell, D. (1991). The practice of science: the research–development relationship with particular reference to agriculture. *Proceedings of the Gadamer Action and Reason Conference*, pp. 25–30. Department of Architecture, University of Sydney.
Fell, L. and Russell, D. (1994). Towards a biological explanation of human understanding. *Cybernetics & Human Knowing*, 2(4), 3–15.
Foerster, H. von (1992). Ethics and second-order cybernetics. *Cybernetics & Human Knowing*, 1(1), 9–19.
Foerster, H. von (1984). *Observing Systems*, 2nd edn. Intersystems Publications, Salinas, California (first edition 1972).
Gates, R., Fell, L., Lynch, J., Adams, D., Barnett, J., Hinch, G., Munro, R. and Davies, I. (1992). The link between behaviour and immunity in sheep. In *Behaviour and Immunity*, ed. A.J. Husband, pp. 23–41. CRC Press, Boca Raton.
Glanville, R. (1995a). A (cybernetic) musing: Control 1. *Cybernetics & Human Knowing*, 3(1), 47–50.
Glanville, R. (1995b). A (cybernetic) musing: Control 2. *Cybernetics & Human Knowing*, 3(2), 43–6.
Glasersfeld, E. von (1985). Declaration of the American Society of Cybernetics. *American Society of Cybernetics Newsletter*, No. 24, pp. 1–4.
Ison, R. (1993). Changing community attitudes. *The Rangeland Journal*, 15, 154–66.
Kövecses, Z. (1990). *Emotion Concepts*. Springer-Verlag, New York.
Krippendorff, K. (1993). Major metaphors of communication and some constructivist reflections on their use. *Cybernetics & Human Knowing*, 2(4), 3–25.
Lakoff, G. and Johnson, M. (1980). *Metaphors We Live By*. University of Chicago Press, Chicago.
Leunig, M. (1990). *The Travelling Leunig*. Penguin Books, Ringwood.
Mahoney, M. (1988). Constructive metatheory: 1. Basic features and historical foundations. *International Journal of Personal Construct Psychology*, 1, 1–35.

Maturana, H. (1988). Reality: The search for objectivity or the quest for a compelling argument. *Irish Journal of Psychology*, **9**, 25–82.

Maturana, H. and Varela, F. (1980). *Autopoiesis and Cognition*. Reidel, Boston.

Maturana, H. and Varela, F. (1987). *The Tree of Knowledge – The Biological Roots of Human Understanding*. New Science Library, Shambala Publications, Boston.

Maturana, H. and Verden-Zöller, G. (1993). *Liebe und Spiel: Die vergessene Grundlage des Menschlichkeit*. Carl Auer Verlag, Berlin.

Mingers, J. (1991). The cognitive theories of Maturana and Varela. *Systems Practice*, **4**, 319–38.

Russell, D. and Ison, R. (1993). The research–development relationship in rangelands: an opportunity for contextual science. *Proceedings of the Fourth International Rangelands Congress*, Montpellier, 1991, Vol. 3, pp. 1047–54.

Van der Vijver, G. (1997). Who is galloping at a narrow path? Conversation with Heinz von Foerster 02/06/1995. *Cybernetics and Human Knowing*, **4**(1), 3–15.

Varela, F. (1979). *Principles of Biological Autonomy*. North Holland, New York.

Varela, F. and Couthino, A. (1991). Second generation immune networks. *Immunology Today*, **12**, 159–66.

Webber, L., Ison, R., Russell, D., Major, P. and Davey, P. (1992). Extending invitations to express enthusiasms for action: a basis for participatory research and development. *Proceedings of the Second World Congress on Action Learning*, pp. 101–104.

Winograd, T. and Flores, F. (1987). *Understanding Computers and Cognition – A New Foundation for Design*. Addison-Wesley Publishing Co., New York.

3 Technology: transforming grazier experience

Raymond L. Ison

3.1 West of the Darling where crows fly backwards

I had not been to the Western Division of New South Wales since I was seven. At that time we were camped near Enngonia, north-west of Bourke where my father was building roads and sinking tanks, large earthen dams, with his dozer and truck and aided by a group of sub-contractors. We were camped in a cluster of caravans around a local shearing shed. I remember stories of mythical shearer's cooks and other characters of the sheds – but no stories of the road builders or tank sinkers. And our pet kangaroo, a baby or joey, the result of an evening's spotlight hunting, an event both enthralling and appalling to a young boy. Also the moving sea of kangaroos that we regularly encountered on the racetrack on our way to town. The only green grass for miles. The Enngonia pub, since burnt down, was the social centre of the district. Its verandah was designed for the dangling legs of a boy, on the edge, ever eager to be part of an adult world. It provided a vantage point for watching every detail over a sarsaparilla and lemonade. Some events moved me from my vantage point – it was around the back for the pig's demise and transformation into ham: the sharp squeal, the blooding, the boiling water, razor sharp knife and scraping.

It was a time of good humour; we had a 1927 Overland car, something of an antique even then but at a time when 'antique' meant 'old'. In other words we could afford no better. We suffered the dust and corrugations of the unpaved roads. The car had to be hand cranked. Experiences were always filtered by the context; distance, dust and with rain, the thick mud which clung to boots, made it impossible for small boys to lift their shoes and for vehicles to move. Rain and mud gave yet another meaning to distance.

Our research project in the Western Division, and planned visit, drew these memories to the surface and shaped my anticipation. We were concerned with the so-called 'failure of graziers to adopt technology' which had been developed by research funded partly with their money. This was a major concern of researchers and extension people, particularly in the local NSW Department of Agriculture. This explained the location of our project. As a group of researchers we had been critical of much of what had been done in agricultural R&D because, as with much of science, it was conducted out of context. We felt it necessary therefore to immerse ourselves in the context of the semi-arid rangelands, where our research was to be conducted. This explained my presence on Murtamena station, not far from Wilcannia, early in December 1990. It was over thirty years since I had been west of the Darling or into 'the outback'.

We, Lynn, my research assistant and I, were full of enthusiasm; we had great expectations. My earlier research in the Forbes Shire in NSW and in the Emmental of Switzerland, had revealed how moving and rewarding the process of actively listening to local people could be. We were a little early on our first visit and it appeared no one was home. Soon in a swirl of dust and from around the corner of one of the sheds appeared two young boys on their motorbikes. Both had been helping with the mustering. I was immediately reminded of a story from a friend who, at the time, had been a kindergarten supervisor with the correspondence school network in Western Queensland. On a round of visits she had been charged with assessing the development of motor-coordination skills amongst her charges. One five-year-old, with parents, when asked to tell whether he could hop on each leg became rather puzzled. No one seemed sure but there was general agreement that he was a great help with the muster, which he had been doing alone on horseback since the age of three. The question about hopping was quickly brushed aside by my friend.

In the midst of making friends with the kids and the family dog we became aware of the approach of some form of mechanised technology, which from the sound, was outside our experience. Children and dog were unconcerned. Then from the sky descended father in what I vaguely knew to be a gyrocopter. If only I had seen the movie 'Mad Max' all would have been clearer. What does 'Mad Max' have to do with it? Well you might ask!! The film had been shot nearby, and the gyrocopter, a key technological feature of the movie, we were told, had captured the imagination and enthusiasm of a local grazier/inventor. He now made them and increasingly they were being used to muster sheep from paddocks which had become too densely covered with invasive 'woody weeds' (Box 3.1) for conventional ground mustering. Was this an example of a grazier who had failed to 'adopt new technology'? I could only marvel at his expertise and despite his enthusiasm, remained steadfastly of the view that it was a technology I had no intention of 'adopting'. We were to discover later that gyrocopters were on the fringe still with most graziers but ultra lights (single person planes) and fixed wing aircraft were commonly used for mustering and checking the network of watering points that kept livestock alive in the unforgiving summer weather – the locals referred to this as doing the bore run, as most of the water came from underground water pumped to the surface with a windmill on a bore.

Box 3.1 Woody weeds in the Western Division of NSW

Several native shrubs from 1 to 4 m high have become a significant feature of the vegetation communities of the semi-arid woodlands in the Western Division of NSW. Some examples are broad- and narrow-leaf hopbush (*Dodonaea* spp.), budda and turpentine (*Eremophilia* spp.) and punty bush and silver cassia (*Cassia* spp.). It is argued that land management practices over the past 150 years have led to an increase in the density of these shrubs, to a stage where they are now regarded as **woody weeds**.

A complex set of factors have given rise to this problem. Many see a main cause as the increase in numbers of stock per hectare that feed on the native pastures. Some blame the current sheep and cattle grazing and the actions of pastoralists, but it is clear that the problem had its origins well before this generation of graziers were born. Overgrazing is related to what is called 'total grazing pressure' (Box 3.3) involving domestic, native and feral animals. Stock eat the pasture, but because the shrubs are unpalatable, they have grown up and crowded out the pasture grasses. Despite the extent of the infestation many current graziers experience the land as being in better condition now than in their parent's generation.

It is estimated that over 12 million hectares is currently suffering from moderate to heavy encroachment by woody weeds, and that a total of 20 million hectares, 70 per cent of western NSW, is susceptible to what is now increasingly described as a serious form of land degradation. A few graziers and researchers question whether this is not a cyclic phenomenon, and argue that the semi-arid lands are more resilient than we appreciate. At the moment, however, they certainly cause hardship for graziers not only because of the loss of grass under the bushes, but because they restrict mustering, forcing many to take to the air to guide those below on motorbikes to the sheep.

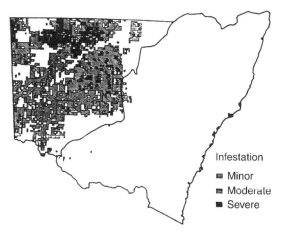

Areas of moderate to severe woody weed infestation in western New South Wales

Source: Adapted from Soil Conservation Service (1992); Harland (1993).

3.2 Why the NSW Western Division?

NSW Agriculture, the state government R&D organisation, was a collaborating organisation in our research. In formulating our project, staff from their western regional office had named 'the failure of technology adoption by graziers in the NSW Western Division' as a problem. They were prepared to be involved in the project if it addressed this problem. At that stage we felt it was important that NSW Agriculture, as an organisation, be involved so we were prepared to accept their statement of the problem as a starting point for our project. Their involvement explained why we were in the Western Division. The language of our project proposal, particularly phrases or metaphors like 'technology transfer', written for the funding agency, the Wool Research & Development Corporation, had little meaning to graziers. They were, however, concerned that research did not seem to be meeting their needs and that they had little say in what research was being done. After all they were paying for half of it. They were prepared to talk with us and to help us with our research when invited. We were from the University and part of the 'agricultural research system', which is not unlike a sort of club, although possibly we were considered by some to be 'radical' members of the club.

Visiting the western Division and talking – really, listening – to local people was the main way we chose to develop our understanding of context. We chose other ways as well. Fiction, poetry and autobiography are also possible ways to appreciate context – means to interpret other's experiences. Earlier in my career I had found this a useful and personally satisfying approach prior to, and during periods of work in Bali (Indonesia), Negros (the Philippines) and Tanzania. I saw no reason why it should not apply to the NSW Western Division, which was equally 'foreign' to me. And of course in this sense all history, except perhaps our own, is 'foreign' to us and we must rely on narrative interpretations.

3.3 Making things easier – transforming the land

When European Australians settled the NSW Western Division they came by river from the south – down the Murray and up the Darling river. They brought with them livestock but for a long time these livestock were limited to the country fronting the major permanent streams (the Darling River) and the country within walking distance of the longer-lasting water holes on semi-permanent streams (the Warrego, Paroo, Bulloo, Diamantina and Georgina Rivers, Coopers and Yancannia Creeks). Of course so too were the native marsupials. Technologies changed all this. Jill Bowen in her biography of the 'cattle king', Sir Sydney Kidman, outlines the sequence of developments. The technology, drilling, was introduced from the USA where it

> **Box 3.2 Bore-capping and the Finlayson Trough**
>
> **Bore-capping**
> On a worldwide basis, only 3% of the useable fresh water is stored on the lands' surface. The remaining 97% is held underground as 'groundwater'. However, in NSW, groundwater makes up only 10% of the 7.2 million megalitres estimated to be used every year. Much of inland Australia is well endowed with groundwater. It is found in underground basins (the most well-known of these is the Great Artesian Basin, which underlies 65% of the state of Queensland) and fractured rock beds. Groundwater in NSW originates mainly from these rock beds of unconsolidated sediments.
>
> Groundwater originates mainly from rain soaking down through the land surface. Most groundwater requires pumping to the surface using either windmills or motor driven pumps. The exception is groundwater held under pressure, for example in the Great Artesian Basin, which flows naturally to the surface through bores and natural springs. Bores are drilled from the surface down to aquifers under strict specifications set down by the Water Resources Commission, to provide water for stock, crops and human consumption.
>
> At present 750 of Queenslands' 2300 flowing artesian bores do so uncontrollably. Many of these bores discharge large volumes of water into watercourses and swamps to be lost through evaporation and seepage. This also provides watering points for animals such as rabbits and kangaroos, which compete with stock for feed (see Box 3.3). The Great Artesian Basin Rehabilitation Program is aimed at reconditioning or 'capping' these bores so that water flows can be controlled by gate valves and used only when needed.
>
> Participation in the programme is voluntary, however, only 20% of the cost of reconditioning a bore is paid by the landholder, and the rest by the government. The total cost of reconditioning a bore ranges from a few hundred dollars up to $40 000 to completely recase and cement a bore. If a new bore is planned, then the old bore must first be completely plugged. So the cycle comes full circle – from governments actively spreading the 'Bore' technology to attempts to control and preserve the underground water today.
>
> **Finlayson Trough**
> Alan Finlayson, a pastoralist from Jeedamya Station, Kalgoorlie, in the semi-arid zone of Western Australia, rigged up a simple device in the 1940s to restrict kangaroos from drinking, whilst still allowing sheep to drink. It was later named

was allowing them to exploit underground water and oil. The first flowing bore was sunk at Kilara station near Bourke, NSW in 1878. As Jill Bowen explains: 'Kidman was fascinated by the process from the outset and excited by further developments which by 1884 saw the NSW government starting to sink bores along roads and stock routes, and with such success that places hitherto impassable through lack of water were now opened up. Kidman had marvelled when the dry track from Wanaaring to Milparinka had been made passable, thanks to this underground water'. At the same time railways and the telegraph penetrated the inland; Thomas Sutcliffe Mort developed refrigeration, making the export of meat possible. These developments transformed the NSW Western Division irrevocably. They shape the nature of the technologies that are being developed today, in an attempt to undo the negative effects of these 'innovations'. Examples

> **Box 3.2** (*continued*)
>
> the Finlayson Trough, and consisted of a low-lying electrified wire surrounding a watering trough at a distance of 1.1 m from the trough's edge. The electrified wire is placed so that sheep would step over the wire to drink, but kangaroos received a shock through their tails and feet.
>
> This first model failed to receive any recognition from the authorities, until John Law from the WA Department of Agriculture carried out preliminary tests (G. Elliott, pers. comm.) No further research was done until recently, in the western division of NSW. The most recent design, the 'Kangaban', appears to be the most effective to deny kangaroos and also allow sheep to avoid being shocked.
>
> Selective watering troughs such as this are not designed to perish kangaroos from thirst, but rather to congregate them over a 6–7 day period to facilitate commercial shooting. Conservation objectives are met because non-electrified troughs outnumber electrified ones. However, there have been mixed reactions to this method of control from the farming community. One response to the Mulga Line, a newsletter put out by the Department of Primary Industries, Queensland, suggested that this was a barbaric practice.
>
>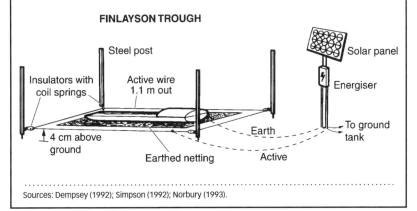
>
> Sources: Dempsey (1992); Simpson (1992); Norbury (1993).

include bore capping in Queensland and the Finlayson trough, which restricts access to water to certain species of animals (see Box 3.2). Aspects of the transforming effects of particular technologies are also explored by Adrian Mackenzie in Chapter 4.

The sinking of bores was accompanied by means to sink earthen tanks, to store water, and to distribute water. Some bores required pumping and with the introduction of steam engines a further transformation of the landscape began. The engines required fuel, which was usually all the available trees in the vicinity of the bore. Together with the insatiable demand for timber in the expanding Broken Hill mines, this accounts for the almost total destruction of tree cover in the area of the Barrier Ranges around Broken Hill.

Concern with the technologies of obtaining and maintaining water supplies remains one of the central preoccupations of graziers today: life for at least six months of the year is organised around the water or bore run. As one grazier reflected:

You could get permanent waters if you had a deep well or bore but ... not tanks or dams that would last longer than a year because you didn't have the machinery to dig them deep enough and the secret to storing water on this country is depth ... you get eight feet of evaporation whether you use any water or not ... it was the advent of machinery during and after the Second World War that you could dig deep tanks, something over 12 feet deep ... before that you had camels and horses ... plus the fact of being able to distribute it ... you could put down steel pipelines but on the advent of poly pipe in [19]55–56 it put a new dimension ... you didn't have to concentrate on a few watering points ... the development of new pumping techniques with mono and multi-stage ... allows you to put water over great distances ... that has led to improvement in the country.

During the summer months pastoralists are tied to their stations because of the need to constantly monitor and maintain their stock watering systems. Failure to do so can lead to heavy stock losses. In the ethics of being a grazier, those who perish a sheep are thought of in critical terms by their peers. To perish sheep shows inattention to waters, or lack of understanding of the behaviour of sheep, both of which are not part of being a sheepman (G. Curran, personal communication). It was the widespread use of polythene pipe or polypipe that to me seemed to have transformed pastoralism and as a consequence the routines of the pastoralists, more than any recent technology. As one grazier noted: 'Pipelines now ... when the tanks go dry I've got pipelines; before I'd have to cart water.' ... 'Polypipe and pumps ... have made things much easier'. Graziers however aspired to be freed from the shackles of the bore run; thus for some there is keen interest in technologies that might provide this release or make it easier for them. For example 'Electronic radio checking (for waters) we'll be going for once it's available'.

3.4 'Built-in' dependence

Our very first visit to Western NSW saw a theme emerge that was to recur again and again. Graziers were greatly concerned about designed-in dependence in almost all technologies that they had to use. This came home to me most vividly towards the end of my involvement with the project. I had spent the day helping with the marking, jetting and mulesing of lambs on one of our collaborators properties. It had been a hard day's work for us all – by all I meant husband, wife, son, eighty-year-old grandfather, who had come up from Adelaide to help at this busy time, as well as the farm hand

and another visitor, from Melbourne, who was 'doing a bit of shooting'. So are most farming operations run in Australia today!!

Work did not stop when we arrived home. It was time for repairs and preparations for the next day. The universal joints on the small Suzuki jeeps were 'floppy' and needed replacement. These farm vehicles held together by ingenious skills of the local graziers usually outlasted the supply of replacement parts. This left only two options – stock up on parts or adapt and bastardise what you had or could get. That night we reconstructed the universal joint from a Holden spare and the damaged Suzuki original. I now understood more clearly why every station we visited had a well-equipped workshop, almost as good as many of the professional mechanics of Sydney. A good hammer was an essential tool for reconstructing the universal, however the hammer was useless when it came to sophisticated electronics and computer chip technology. This was their concern. No longer could they adapt, innovate and sustain their operation from their workshop. Even the local inventor, or increasingly, the regional specialists a day or more away in say Broken Hill, lacked the skills or the parts to meet their needs. They were increasingly locked into a technology web with its epicentre in Osaka or Silicon Valley and which came to them via Adelaide or Sydney on the midday flight only after the problem had, hopefully correctly, been identified, the order placed and the supply location tracked down. Independence and autonomy were prized manners of living west of the Darling; inexorably technology was taking these prizes away.

Dependence was not only in the form of things – spare parts, or new machines, but also on people, particularly service people. The debate about grid power installation, which was dividing the community during our period there, exemplifies the situation. Ironically we were exposed to both perspectives on the one day. At the first property we became aware of the enthusiasm the owners had for their solar-powered system, which they had devoted a lot of time and effort to developing and installing. Despite slightly higher costs they saw this as the future, a future without the blight of power lines defacing their landscape and a future which preserved their commitment to self-sufficiency, local autonomy, and environmental friendliness. They had also placed a lot of effort into ensuring the resilience of their system. They maintained appropriate spares but also an entire backup power system based on the more common traditional diesel engine. This system was also testimony to what could be achieved by way of innovation by individuals pursuing their enthusiasm and identifying the necessary people and data to translate their enthusiasm into action and desired outcomes.

At the very next station the same topic came into the conversation, a conversation held to the persistent thump, thump . . . of the diesel power system. The contrast with the previous household made it more apparent. This family, the wife in particular, were determined to be connected to the new power grid. They yearned to be free of the uncertainty and unreliability of their current system and the increasing reliance they felt on service personnel to repair their system or appliances affected by inadequacies in their system: 'You can't rely on them. They never come when they say they will!' Their dream was that with the flick of a switch they would no longer be dependent. Similar concerns, but different interpretations of experience and thus different strategy. This points to the need to provide variety.

During our interviews it became apparent (to us as observers) that the graziers had many experiences of a similar nature, but that their interpretations of, and emotional predispositions to, issues and events were expressed in many different forms. This resulted in a diversity of actions. For example, we encountered another grazier who was enthusiastically developing his own solar power system. He explained how his enthusiasm had been triggered:

'One thing that convinced me about solar was that we used to have a boiler outside. I used to pay the kids to light it, but after the kids left we got stuck with it ourselves.'

It was his view that:

'For one-third of the cost you can be self-sufficient using sun and wind.'

and that:

'We'll go to a fair amount of solar to cut power. It will be economical and a better way of doing it than just generator.'

Others explained why they felt the proposed electricity scheme would not deliver the anticipated benefits:

'I wanted to do it but saw it was too expensive – $80 000 and still subject to power failure.'
'They are installing the power lines around Menindee Lakes, but have put the poles too far apart and the lines break in the wind. So you need a generator back up anyway. They might have sorted it out by the time it gets to us I hope.'
'How do you tell people, they've got no money to feed their kids, pay their debts that they should be putting up a power line . . . when we know in 20–25 years time it might be the best thing that's ever done.'

These stories give us glimpses of the diversity of actions which arise in particular contexts around what seems to be a common issue. Whilst the issue is common, what is different are the individual experiences which give rise to particular behaviours, and even in this relatively small community a great diversity of behaviours is apparent. This is the crux of our dilemma. How are the diversity of people who constitute the 'grazier' or the 'researcher', or who together will constitute the future 'rangelands', to participate in the networks that give rise to new technologies, whether they be things, practices or policies?

3.5 But the land is better now than when I took over from my father!

We listened to many graziers in the Western Division. A common theme in what we heard, particularly from those with a longer history in the area, was that the land was now better than it had been in their father's time. As one grazier said:

Where a lot of the catch-cry is the Western Division is degraded . . . in the last 20 or 30 years . . . when the rabbits were knocked down after the myxo . . . in 1950–51, the advent of poly-pipe, better management of paddocks and smaller numbers . . . because you could put more waters round you could split your stocking numbers down; you've got steel fabricated fencing and more mechanisation . . . people tend to get more mechanised to do the work. Things are better now.

Their experience was that the land was now in better shape than it had been. In contrast, the prevailing discourse in scientific or R&D circles was that the land was becoming increasingly degraded. This was another version of the Botswanan story told by Louise Fortmann and outlined in Chapter 2. Graziers in the Western Division were not conscious of having failed to 'adopt technology' – they did not see this as their problem. They were aware that research was being done on burning as a means of vegetation management, also blowfly control and that some had set up grazing experiments in the past although they remained very sceptical about extrapolation from small plot experiments. When pushed they would admit to having little confidence in particular research, of not trusting someone's advice because of a particular bad experience or acknowledging that so and so was 'a good bloke' but 'I didn't realise he worked for the Ag department'. By and large the concerns of researchers and those who fund R&D were not the concerns of graziers and most graziers knew little about the organisations and people involved.

3.6 If the dog won't eat it, it ain't dog food!

As we became more immersed in our understanding of the different discourses in the Western Division, it became apparent that for many researchers the so-called problem of technology adoption was related to one issue – undesirable changes in the vegetation of the Western division leading to loss of productive capacity. This was mainly manifest in the so-called 'woody weed' problem (see Box 3.1) and in research that pointed to loss of perennial grasses as a percentage of the total vegetation. Most graziers and researchers recognised this as a complex set of interrelated issues but there was not universal agreement that it was a 'problem'. Contrary to the mainstream researcher view, some graziers saw positive effects from the soil binding and erosion control, offered by the woody shrubs – preferring this term as opposed to 'weeds'. Three specific technologies were most often spoken about in R&D circles and had been the focus of research effort. These were (i) burning; (ii) stock management, particularly, stock numbers and (iii) techniques for 'control' of the 'woody weeds'.

The so-called technologies of burning illustrate some aspects of the situation. Most graziers appreciated that the vegetation had evolved under regimes of fire. But things were now different. Most appreciated that if you locked up a paddock long enough and allowed a good growth of fuel to get a hot burn then you could help control woody weeds and possibly encourage perennial grasses. But who was going to take responsibility for lighting the match? They faced the possibility that they might be sued by their neighbour if a fire they had lit got out of control. This had obvious financial and social implications. There were also trade-offs to be made – locking up paddocks meant production and income forgone, even if in the long-term there might be something to be gained. As much as they wanted to, it was often hard to think in terms of the long haul when you were in the middle of a roaring drought, wool prices had collapsed, there was a need and high priority placed on educating your kids (requiring about $15 000 per annum per child in secondary or tertiary education), or because the tax system rewarded you for 'spending up big' in good years on things that might not be really necessary. For some, who had followed the commonly articulated expert advice in the 1980's of 'get big or get out' the situation was even more desperate because of the high interest rates combined with low wool prices. To make matters worse their 'landlord', the Western Lands Commission would not allow them to diversify outside their designated landuse of grazing, which precluded on-farm tourism and the like.

Over the period of our research it became apparent to us that all the technologies available for woody weed control were uneconomic, something that most graziers, concerned though they might be, had recognised for some time. This left the thorny issue of stock management and particularly stock numbers, which had been named as an issue of concern by institutionalised authority since at least the 1880s (see Box 3 and Chapter 4).

> **Box 3.3 Total Grazing Pressure and the management of feral animals**
>
> In the semi-arid rangeland districts of Western NSW, where stock are run on native pastures, managing the Total Grazing Pressure (numbers of ALL animals that feed off the pasture) is essential. By matching the number of animals to the amount of feed available, many short-term feed droughts can be averted. In the past, however, this matching has not occurred on many properties, and stocking has been maintained or maximised regardless of the available pasture. This has led to major feed deficits which result in malnutrition, poisoning from eating toxic plants (not normally eaten when the pasture is in good condition), and starvation.
>
> There are also other consequences of overgrazing. As the pasture species are eaten out, woody weeds grow in dense stands see Box 3.1), pasture does not regenerate, soil structure is weakened, increasing the risk of erosion, and there is no fuel available to burn the woody weeds.
>
> Careful monitoring of numbers of stock and other herbivores such as rabbits, goats and kangaroos, and taking the appropriate action, will avoid these problems.
>
> When managing the total grazing pressure, the number of stock (stocking rate) placed in a paddock is determined by the availability of water, the value of the pasture species available, the accessibility of the pasture to the stock, the cover of trees and shrubs and the number of other herbivorous species present.
>
> Rabbits are the most widespread and abundant of vertebrate pests of the rangelands. Removal of rabbits greatly increases the ground cover and the regeneration of valuable pasture species. Due to the vastness of the district, many control methods are not economical, and the most recommended means of control is mechanical ripping of the burrows.
>
> Kangaroos have increased in abundance in western NSW due to the availability of permanent water supplies. The only legal forms of control are fencing (expensive) and shooting.
>
> Feral goats and pigs are also a problem to landholders. Because the diet of sheep and goats overlap, they compete directly with sheep for food. Goats can be trapped and shot. Feral pigs are the major predator of lambs in western NSW, and can spread disease. Trapping, poisoning and shooting are most effective in the dry season, and protect the newly born lambs.
>
>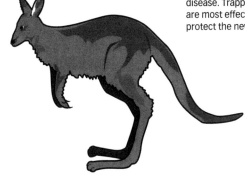
>
> Source: Adapted from Curran (1992); Curran et al. (1993)

Whilst there is no doubt that some graziers were or had been overstocked – in fact graziers were prepared to talk quietly about this, and were the main source of complaints about overstocking to the Western Lands Commission, – what is often not appreciated is the complexity of this issue both biologically and socially. Too frequently there has been a tendency to blame rather than to turn to dialogue, although when all is said and done this is the only way in which the issue can be dealt with. Despite legislation designed to give force to land management issues, the accepted understanding in official circles is that rarely would attempted prosecutions stand up in a court of law.

This focus of institutionalised authority on graziers does not seem to have been a new phenomenon. Earlier this century the same dilemma was apparent but the players in the drama were cast in contrasting roles. This is well illustrated in Jill Bowen's biography of Kidman in a description of a debate in the South Australian Parliament in 1916 – the then Minister of Industry stated: 'Before Mr Kidman came in, there was a greater number of cattle and sheep in South Australia. He must stock the land and use it properly or give it up' (Bowen, 1987, p. 234).

3.7 Technology – a mediator of experience

What does technology adoption or technology transfer mean? Generally in agricultural R&D a technology is thought of as a product, such as a particular dip or drench, or a set of practices, such as regular burning to control woody weeds. It is not generally recognised that organisations, because they embody sets of practices, might be considered as technologies as well. A common definition given to technology is: *'the application of scientific knowledge to practical tasks'*. From this perspective science is generally seen as the only source of new knowledge which becomes embodied in a tool, machine or technique – an artefact. Few would question the capacity of science to explain phenomena. This is its strength. In practice those who do science, or find themselves part of the R&D system, do not usually stop at explanation. This gives rise to the model common within agricultural R&D:

Science/research → knowledge → technologies → transfer → adoption (by farmers) → diffusion (amongst farmers)

This particular model has increasingly been found wanting as an explanation of what occurs in the conduct of science and R&D. From studies of the sociology of science have come understandings that technology often precedes science – the classic example is the development of lenses from craft knowledge, which enabled the development of the telescope, and thus the observations of Galileo, from which his scientific understandings were derived. Technology thus gives rise to new questions for which we seek explanations. This has led some observers to speak of technoscience rather than science and technology.

When researchers move beyond explanation, into application or 'transfer of knowledge or technologies', they immediately move into a sphere of work where they are attempting to persuade or to enlist people to do something –

ultimately to change their behaviours in a way that is seen as desirable by the researchers, or the system of which they are a part. In the case of agriculture this process changes both farms and the lives of those who farm or depend on the products of farming. Some people describe these as 'technological impacts' whilst others reject this metaphor in favour of the notion that technologies amplify or suppress, or reveal and conceal, particular aspects of the ways in which we experience or perceive the world. Agricultural technologies, for example, are designed to amplify certain productive processes just as, say, the fork-lift, is designed to amplify our ability to lift. Of course just as technologies might amplify so too might they suppress, as for example our sense of touch, or of physical exertion with the fork-lift example. The notions of amplifying/suppressing or revealing/concealing tell us something about the nature of the relationship between us, a technology and 'our worlds'.

This gives rise to an alternative definition of technology as a selective amplifier. As the philosopher Cliff Hooker observes: 'What it amplifies is a function of its own nature and design (and just possibly a function of the system into which it is inserted). What effects the amplification has depends on the system into which it is inserted. A human–technology system is a mutually adaptive one, evolving a set of lifestyles and technological designs in successive interactions'.

3.8 Technology diffusion or building networks?

One of the main theories that has shaped the practice of agricultural extension in the last thirty years has been the 'diffusion of innovations model'. As outlined in Chapter 2, at its simplest extension officers were seen as being involved in transferring knowledge, usually in the form of a practice or set of practices, in an attempt to get graziers to adopt or use that knowledge. The diffusion of innovations model developed, particularly by US rural sociologists, was then employed to explain the adoption or non-adoption process. The diffusion and adoption of innovations model of technology transfer, with its associated language (Figure 3.1) shaped how extension is thought and talked about, especially amongst administrators from a research background. As outlined earlier, it was because staff of NSW Agriculture did not believe that graziers were adopting or using particular technologies that we had come to the Western division in the first place. This raises the question of why staff of NSW Agriculture had named the problem in this way. This is explored further in Chapter 5.

In preparing this chapter I had occasion to re-read a book which has been very influential in shaping how agriculturalists and veterinarians think about extension or technology transfer. In the preface to his first book *The Diffusion*

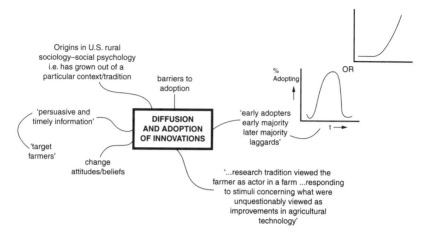

Figure 3.1
The language associated with the diffusion and adoption of innovations model of technology transfer and some of its questionable assumptions for rangeland R&D. (Adapted from Buttel *et al.*, 1990.)

of Innovations, Everett Rogers describes a key experience which shaped his subsequent lifetime commitment to diffusion research – scholarship which I might add has shaped generations of agricultural students understandings of what it is to do 'extension'. He described it thus (p. ix): 'My interest in the diffusion of innovations began when I was an Iowa farm boy. After high school and college training in modern agriculture, I found my home community somewhat less than impressed with my stock of innovations'.

I do not think Everett Roger's experience was unique although his response to it may have been. In my experience it is common for those involved in professional practice to name a problem as that of their clients not doing what they, the so called 'expert' thinks is good for them – of questioning how well their clients have carried out the 'expert' advice which they have given. My own first experience of this occurred whilst still an undergraduate when my father, a grazier, was the focus of my advice about pasture improvement. Choice of species and method of establishment were not problematic and accomplished as per instructions. However, pastures have a potentially long life in contrast to the period in which a young son is present or interested. In providing advice about species and sowing I was of course also envisioning, but not articulating, future states based on my *theories* of grazing management and factors governing pasture persistence. These were products of my experience but of course not of my father's. My father was not satisfied with the outcome. I of course replied: but you did not do x and then y!

The diffusion of innovations model was itself derived in a particular context, mainly in North America and Europe, based on some well-known case studies such as the 'adoption' and 'diffusion' of hybrid corn in the USA. As so often seems to happen, educators, researchers and administrators have sought to employ this model in contexts in which its assumptions no longer hold. This has been known for a long time, even in an Australian rangelands context, yet still it shapes people's perceptions of what extension is (see Chapter 5). Rogers has come to recognise this in part himself. In the preface to his third edition of *Diffusion of Innovations* he acknowledges that 'many diffusion scholars have conceptualised the diffusion process as one-way persuasion' and that 'most past diffusion studies have been based upon a linear model of communication defined as the process by which messages are transferred from a source to a receiver'. Rogers still has a focus, however, on 'information transfer' based on the work of Shannon and Weaver (1949); we have questioned these assumptions in Chapters 1 and 2.

A much more robust critique of the 'diffusion of innovations' model has been offered by Bruno Latour in his work *Science in Action. How to Follow Scientists and Engineers Through Society* (1987). As the title suggests his arguments derive from studying what scientists do. From this he observes that 'facts' are constructed in networks and embodied in what he calls 'black boxes'. We might see these black boxes as incipient technologies. Thus (p. 132): 'claims become well-established facts and prototypes are turned into routinely used pieces of equipment. Since the claim is believed by one more person, the product bought by one more customer, the argument incorporated in one more article or textbook, the black box encapsulated in one more engine, they spread in time and space'. It is this spread in time and space that has been the concern of diffusionists; as Latour observes: . . . 'If everything goes well it begins to look as if the black boxes were effortlessly gliding through space as a result of their own impetus, that they were becoming durable by their own inner strength. In the end, if everything goes really well, it seems as if there are facts and machines spreading through minds, factories and households, slowed down only in a handful of far-flung countries and by a few dimwits'. The diffusionist model is captured in common metaphors such as the 'spread of ideas', 'the march of science', 'the irreversible progress of science' or the 'irresistible power of technology': 'first, it seems that as people so easily agree to transmit the object, it is the object itself that forces them to assent. It then seems that the behaviour of people is *caused* by the diffusion of facts and machines. It is forgotten that the obedient behaviour of people is what turns the claims into facts and machines; the careful strategies that give the object the contours that will provide assent are also forgotten . . . the model of diffusion invents a technical determinism, paralleled by a scientific determinism ... Facts now ... seem to move without people. More fantastic, it seems they would have existed without people at all'.

From Latour's perspective the diffusion model and its associated metaphors are an inadequate description of his experience of studying scientists at work. Not only are facts seen to move of their own accord from the diffusionist perspective, but they are seen to reproduce themselves. This fails to account for novelty – our experience that facts and technologies are constantly changing and are not simply reproduced. Facts or technologies are, in the diffusionist model, said to have originators who can be discerned by careful study of genealogies of 'technical descent' and who are made 'so big that they now have the strength of giants with which to propel these things!' Rarely would this seem to be the case. These great inventors, 'men and women of science are always a few names in a crowd'. As Latour observes (p. 135): 'It is ironic to see that the ideas which are so valued when people talk of science and technology, are a trick to get away from the absurd consequences of the diffusion model, and to explain away how it is that the few people who did everything nevertheless did so little'. Adherents to the diffusion model hang on to their claims through the language they have evolved to explain away interruptions, deflections, corruptions or just plain ignoring the 'flow of facts': some groups are said to 'resist', or there are 'barriers' to adoption. Thus the task of advisers or extensionists within the diffusion model is to identify and break down the barriers – this is often associated with refining and targeting of the message! (Figure 3.2). I think it is fair to say that the diffusionist model shaped the understanding of most of the people involved in NSW Agriculture and the R&D funding organisations at the time we commenced our project in the Western Division; indeed it still does. Undoubtedly one of the aspirations for our project, by some, was that we would develop new ways in which to get the graziers to do what we, or the R&D community, considered was in their best interests by identifying the so-called barriers to adoption.

The incorporation of the 'diffusion' model in the practices of Departments of Agriculture, and the teachings of university curricula and the like, is in itself a good example of institutions as technologies. One can reflect also on the work that has been done to develop and perpetuate the model as a 'black–box' in which particular relations between different groupings of people are established. Latour's ideas thus enable both a critique of the internal coherence of the diffusion model as well as explaining why it is so pervasive in the NSW Western Division and in the agricultural R&D institutions.

In proposing an alternative model Latour speaks of the model of transla-

Figure 3.2
'Targeting' of graziers or clients in research programmes is a commonly used metaphor. (From *Everyday Evaluation on the Run* by Yoland Wadsworth, 1991, with permission from Allen & Unwin.)

tion in which there is a need to symmetrically consider the efforts to enrol and control human and non-human resources. He poses a question relevant to our experience of graziers in the Western Division: 'How simple is a simple customer?' Even the most sophisticated 'black box' usually requires some attention by the customer or ultimately it fails. In addition, as with the example of the sophisticated electronic 'black-boxes' which graziers found increasingly difficult to adapt or modify, there was still a need to do a lot of work to set up and maintain the system of production, transport and supply, that was capable eventually of getting the spare part from Osaka to the graziers workshop. As Latour observes, 'the more automatic and blacker the box is, the more it has to be *accompanied* by people . . . By itself it has no inertia'. This raises questions of what are appropriate technologies in the context of the sparsely populated Australian semi-arid rangelands and whether enough work has been done in the process of developing particular technologies. Appropriate in this sense means meeting some experienced need as in the case of the kids, who were paid to put wood in the boiler, leaving home. Of course the graziers and those committed to a future wool industry in the rangelands might well ponder on who is going to do the work – particularly if services in rural towns and cities run down. I would suggest that the graziers themselves and people in the local towns will need to be enlisted.

Latour argues that 'we are never confronted with science, technology and society, but with a gamut of weaker and stronger associations; thus understanding what facts and machines are is the same task as understanding who people are' and that 'irrationality is always an accusation made by someone building a network over someone else who stands in the way'. For many researchers, graziers stand in the way of building their new networks. When considered with the following two chapters (Chapters 4 and 5), Latour's proposals give us new insights into, and interpretations of, our experiences with graziers in the NSW Western Division.

3.9 People in networks

Latour's idea that translation is always accompanied by people allows a brief link back to the cognitive theories of Maturana and Varela (1992; see Chapters 1 and 2). The model of translation can be seen as a web which encompasses or enmeshes people, machines (artefacts) and facts. Within this web, technologies, whether they be things, practices, or institutions, shape or filter human perceptions in given contexts.

As Don Ihde (1990) notes, there is a peculiarity with respect to the user of an artefact: the tool or equipment becomes the means of the experience, not the object of the experience. The horizon of our perceptions is not limited by the outline of the human body. Perception at a distance is mediated through an artefact (technology or theory). Distance and proximity as well as history are important when considering perception. This became apparent in a story related to us by one grazier: he told of buying a small miners' cottage in Broken Hill so that his wife and children could move to town to have access to education. So as to be with them some of the time and continue his work he used to save up all the repair and maintenance work and take it into town on his truck. However, working there and absorbed as he was, he often found he would go in search of a tool only to look up and discover that he was three or four backyards away from his own home. This man's perception of distance arose in his working patterns, which had previously been shaped in a different set of relationships with his tools and working space.

Here, following Maturana, we might observe that the history of structural coupling, the relationship of any organism to its environment, is unique to each individual. It is this awareness which enables us to think about the many and creative ways in which invitations to participate in co-creating our future worlds might be extended. Being aware that technology both mediates and extends our perception and transforms our experience also provides new horizons which surpass those embodied in the pervasive diffusion of innovations model.

3.10 Has institutionalised R&D transformed graziers lives?

The Australian poet, Henry Lawson, in his poem 'Bourke', named after a town in the NSW Western Division, relates his experiences of the infamous drought of the 1890s:

No sign that grass ever grew in scrubs that blazed beneath the sun; The plains were dust in Ninety-two, and hard as bricks in Ninety-one. On glaring iron-roofs of Bourke the scorching, blinding sandstorms blew, No hint of beauty lingered there in Ninety-one and Ninety-two.

One hundred years later the newspaper headlines exclaim:

Dirt poor. You think you have heard it all before . . . about how Australia's farmers are suffering. But not on this scale, not in this way. The vast plains of our state's west have become a disaster area – for both people and their properties. Hopes and dreams are being eroded, just as surely as the land itself.

(Wahlquist 1992).

So what has changed? Has institutionalised R&D made any difference and is 'the failure of technology adoption' the 'problem'? In an address to the Australian Rangeland Society Conference in 1992 (see Ison, 1993) I posed these same questions. Nothing said there or which has happened since leads me to conclude that things have changed for the better because of institutionalised R&D or that the so-called failure of technology adoption is 'the problem'.

During our conversations with graziers many technologies were mentioned which had transformed their lives. These included, in addition to those already discussed, aerial technology, drenches, vaccines, jetting races, wide combs, mobile yards, mobile crutching units, black polythene pipe, motorbikes, the network of sealed roads which had led to the demise of local community groups, using computers for budgeting/management and faxes. Whilst institutionalised agricultural R&D has contributed to the development of drenches, vaccines and computer budgeting, the transforming effects of these technologies have been minimal (they are examples of first-order change) in contrast to those of water and roads. Other examples of supposedly recent technologies developed by R&D were sometimes found on stations where they had been in use for many years. Two examples were bugle races and the adaptation of tractor front-end loaders for woody weed control. Clearly graziers are experimenters and researchers in their own right. As Don Ihde has observed, design, in the history of technology, usually falls into the background of a multiplicity of uses, few of which were intended at the outset. There are 'things in context' and contexts are multiple.

The properties we visited were often a man and woman operation covering over 40 000 ha. This is feasible when using aerial technology. For mustering sheep the husband would be in the air and the wife would be on the

> **Box 3.4 'Time's Up. Are You Extending?'**
>
> **A remote country town prepares to say goodbye to one of Australia's few remaining manual telephone exchanges, and a way of life**
>
> Some people call Gloria Jackson the de facto mayor of Ivanhoe because she knows everyone in town and can get things done. Ivanhoe, a remote town of 500 on the far western plains of New South Wales, has a pub, a school, a small hospital, a general store, a cafe, an RSL club and two service stations. It also has one of Australia's last manual telephone exchanges, which gives Jackson and five other telephonists a key role in the town's life. But after almost a century, the exchange is about to close down, taking with it the women's jobs and a colourful slice of Australian rural history.
>
> Jackson, 39, began training as a telephonist in Ivanhoe when she was 17. The little town then had a thriving social life centred on cafes, a cinema that showed movies twice a week and woolshed dances attended by young people employed on the area's huge sheep stations. The cinema and most of the cafes have since closed, and few stations hire permanent workers nowadays. Ivanhoe lost another link with its past recently when the NSW government abolished one of the state's best known passenger trains, the Silver City Comet from Broken Hill to Sydney, which passed through Ivanhoe.
>
> The passing of the telephone exchange will probably leave the biggest gap of all. It services 170 homes in town and another 130 surrounding properties are linked to town on the same line. Jackson and her fellow telephonists keep the exchange running 24 hours a day on four rotating shifts.
>
> Jackson left the exchange after her apprenticeship to get married and have children, but she has been back at the switchboard for 13 years. Not much can happen in the town without her knowing. 'People don't think twice about ringing in and asking us to pass around messages, when someone is sick or a baby has been born', she says.
>
> 'Farmers always ring us first when there's a bushfire, and ask us to warn their neighbours. An electricity linesman rang me one day when it was 48 degrees, and everything was melting, and asked me to ring 30 homes to tell them their power was going to be turned off. Some kids ring us and simply ask to be put through to "Nanna" or "Mum" and we usually know who without needing the numbers'.
>
> It is a hot Saturday afternoon and Jackson is doing the early evening shift on the exchange. She speaks between tending to calls with familiar refrains such as 'Ivanhoe', 'Wait on please' and 'Three minutes. Are you extending?' She is interrupted by a call from the police at Hay, 320 kilometres away, asking about the

ground on the motorbike. This has been one of the most profound transformations in the NSW Western Division – the gender politics of daily work. It is where institutionalised R&D has been years behind the day-to-day experiences of graziers and where institutionalised stereotypes and behaviours have perpetuated old myths. This demonstrates that every technical change is simultaneously and necessarily a social change; to put new technology into practice, you must change behaviour (See Box 3.4).

Graziers of course are constantly engaged in reflection on, and transformation of, their own experience. As with all of us, graziers interpret their experience from their traditions (literally their pre-understandings). They have much more to contribute to the conversations which shape design of future R&D than is currently the case. For all of us though, it is difficult to stand outside our traditions of interpretation. In the design of future forms for the semi-arid rangelands there is a need for a diversity of traditions

Box 3.4 (continued)

state of the unsealed Cobb highway, named after the old Cobb and Co. coaches, between Ivanhoe and Wilcannia. The telephonists are inundated by such calls whenever it rains.

'Once the party lines are gone, people will be even more remote.'

The pattern of work at Ivanhoe has not changed for generations. Jackson and her colleagues time each call with old-fashioned clocks, then write down the details on dockets. They compile the accounts each month, then send them to the Telecom Australia regional office in Narrandera, which eventually sends out the bills. 'Sometimes people query their bills, and ask if they can see the dockets', says Jackson. 'There was a Greek cafe owner in town once who used to speak to Greece for half an hour at a time, then hit the roof when his bill came. We'd just say, "Sorry mate, here's your docket"'.

When the Ivanhoe exchange closes next month, such personal touches will have gone for good. The 150-year-old Magneto manual exchange will go to a museum, and Jackson and her colleagues will be replaced by a big antenna and a small shed containing a computer on the outskirts of town. The old single-digit numbers will become six digits, and Ivanhoe will be linked to the rest of Australia's automatic system as part of Telecom's rural area conversion programme, which began six years ago. Only about 20 remote locations like Ivanhoe remain manual, and all are expected to be converted by 1993.

A few Ivanhoe farmers, such as Jim and Prue Graham, have jumped the gun and already installed their own high-tech. remote control telephones, which enable them to dial the world direct. But Gloria Jackson reckons the changes will take away something essential from the bush's character. 'With the old party lines, every one can talk to their neighbours–and listen in on their neighbours-without ringing the exchange. The party lines are a way of keeping in touch when your nearest neighbour can be 100 km away. Once they're gone, people will be even more remote from each other in a way'.

They are planning a big party in Ivanhoe to mark the exchange's closing. Jackson will remain a Telecom employee, but only because she is moving to Wollongong where she will join a vast, fully automated team servicing reverse charge calls and fault inquiries. 'I'll miss the contact with the people in Ivanhoe most of all', she says. 'At Wollongong, once you get the numbers up on a screen, they all disappear into a big pool'.

From Milliken (1991), with permission.

---

joined in mutually appreciative dialogue (see Kersten, 1995). Such attempts may be enhanced by appreciating that for every revealing transformation there is a simultaneously concealing transformation of the world which is given through a technological mediation. Technologies, including philosophical and scientific theories, transform experience and are therefore non-neutral. As Maturana (1988) observed:

> ... *a source of problems in human relations is our use of philosophical or scientific theories to justify our attempt to force others to do what they do not want to do under the claim that those theories prove that we are right or know the truth while they are wrong or ignorant. And what is worst, we may be sincere in believing our claim because in our ignorance we do not understand what philosophical and scientific theories do.*

3.11 Historical patterns and technological lineages

In shaping our research we placed considerable importance on the notion of pattern – concerns with pattern and form are at the heart of the new sciences of complexity, but they are also at the heart of our notion of streams of tradition (see the introduction to this section and Capra, 1996). Traditional research within the first-order tradition has placed considerable importance on understanding patterns of data – pattern analysis is common in ecology and informs agroecosystem analysis (Conway, 1985). The initial visits we made to the semi-arid rangelands, plus available secondary data, convinced us that little would be achieved by the development of a comprehensive ecological database from which to explore ecological patterns. Others were doing this. Instead we commissioned research to explore, using pattern analysis as a guiding principle, the historical and philosophical patterns of interpretation of R&D action in the rangelands. We wanted to better understand how the rangelands that we experienced had come to be. This research forms the basis of Chapter 4 in the next section.

References

Bowen, J. (1987). *Kidman, the Forgotten King*. Angus and Robertson, Sydney. 476pp.
Capra, F. (1996) *The Web of Life. A New Synthesis of Mind and Matter*. HarperCollins, London.
Conway, G.R. (1985) Agroecosystem analysis. *Agricultural Administration*, **20**, 31–55.
Cunningham, G.H., Mulham, W.E., Millthorpe, P.L. and Leigh, J.H. (1992). *Plants of Western N.S.W*. Inkata, Melbourne.
Curran, G. (1992). Grazing strategy in rangelands: what is grazing strategy, and can strategy be changed? *Rangelands Management Workshop*, Fowlers Gap via Broken Hill.
Curran, G., Lugton, I.W., Davis, E.O. and Jubb, K.F. (1993). Managing the causes and effects of drought. *Proceedings of the Australian Veterinary Association Conference*. Gold Coast.
Dempsey, N.O. (1992). Kangaroo control that works. In *The Mulga Line*, November, p. 7. Department of Primary Industries, Charleville.
Fortmann, L. (1989). Peasant and official views of rangeland use in Botswana. *Land Use Policy*, July, 197–202.
Harland, R. (1993). *Managing for Woody Weed Control in Western NSW*. NSW Agriculture.
Hooker, C.A. (1993). Value and system: notes toward the definition of agri-culture. *Proceedings of the Centenary Conference, Agriculture and Human Values*. UWS-Hawkesbury. Unpublished monograph.
Ihde, D. (1990) *Technology and the Lifeworld. From Garden to Earth*. Indiana University Press, Bloomington.
Ison, R.L. (1993). Changing community attitudes. *The Rangeland Journal*, **15**(1), 154–66.
Kersten, S. (1995). In search of dialogue: vegetation management in western New South Wales, Australia. Unpublished PhD Thesis, University of Sydney.
Latour, B. (1987). *Science in Action: How to Follow Scientists and Engineers Through Society*. Open University Press, Milton Keynes.
Lawson, H. (1990). *Poetical Works of Henry Lawson*. Angus and Robertson, Sydney.

Mackenzie, A. (1992). An analysis of the major technological innovations developed and/or promoted by research and regulatory bodies concerned with management and utilisation of the Australian sheep-grazed rangelands with special reference to western NSW. Unpublished Report, The University of Sydney, 34 pp.

Maturana, H.R. (1988). Reality: the search for objectivity or the quest for a compelling argument. *Irish Journal of Psychology*, **9**, 25–82.

Maturana, H. and Varela, F. (1992). *The Tree of Knowledge. The Biological Roots of Human Understanding*, 2nd Edn. New Science Library, Shambala Publications, Boston.

Milliken, R. (1991). Time's up. Are you extending? *Time Magazine Australia*, Jan, p. 5.

Norbury, G. (1993). Finlayson trough as a means of kangaroo control. *Rangeland Management Newsletter*, Feb., pp. 3–4.

Rogers E. (1962). *Diffusion of Innovations*. The Free Press, Glencoe, Ill.

Shannon, C. and Weaver, W. (1949). *The Mathematical Theory of Communication*. University of Illinios Press, Urbana.

Simpson, I. (1992). *Rangeland Management in Western NSW*. NSW Agriculture.

Soil Conservation Service. (1992). *Total Grazing Management*. Soil Conservation Service of NSW.

Wahlquist, A. (1992). Dirt poor. *Sydney Morning Herald*, June 13, p. 35.

Part II
Historical Patterns, Technological Lineages and the Emergence of Institutionalised Research and Development

The next two chapters are concerned with two projects designed specifically to develop a second-order appreciation of our overall research context. They were, firstly, a pattern analysis of the major technological innovations developed by institutionalised R&D (Chapter 4). Adrian Mackenzie, then a PhD student in philosophy at the University of Sydney, but with an initial science degree, was commissioned by us to conduct a pattern analysis to explore the historical and philosophical patterns of interpretation of R&D action in the rangelands.

Given that the 'problem' had been defined as the failure of 'technology adoption' we felt it important to understand what technological lineages could be identified in the social construction of the rangelands and what role, if any, the organisations set up to do R&D had. We were also attempting to see whether historically there was anything new in the way the problem had been named.

In commissioning the research we had been interested in the notion of hermeneutic patterns, that is how could we interpret the historical nature of the interactions between insiders (pastoralists) and outsiders (the first European explorers; state government bureaucrats and legislators particularly in the late nineteenth century, and more recently agency personnel) in the conduct of R&D? We conceived of a form of textual hermeneutic analysis (interpretation based on our conceptions of R&D) to reveal patterns of perceptions about R&D 'problems' and 'opportunities' and how this may or may not have shaped actions (settlement, land tenure, subdivision, organisation formation and technical innovation). The method employed by McKenzie was to review, using a sampling procedure, all the published material from 1830 to the present. The theoretical framework he chose to conduct his analysis was that of the French philosopher/historian Michel Foucault, whose major concerns have been with discourse and power. Technological innovations raise questions of power because they are one of the means by which power has a material impact on the lives of people: not power in the strictly limited sense of political influences or the means to repress and restrain, but power as what shapes people's capacities to act and even constitutes the modes of action available to them.

The material McKenzie focused on was:

(i) Some of the diaries of the early explorers.
(ii) All select committee reports of the NSW parliament.
(iii) Annual reports of relevant current government bodies (and their progenitors), which particularly included 'service' organisations such as NSW Agriculture, the NSW State Government research and extension organisation, the Soil Conservation Service (another State Government body concerned with both soil and vegetation management in the rangelands) and the Western Lands Commission, the body responsible for implementing and monitoring the legislation contained in the Western Lands Act. These were sampled at ten-year intervals except for the decade 1981–90, which was examined in detail.
(iv) R&D Corporations (meat, wool, etc., one of which was funding our own research) that had allocated any resources to R&D in the rangelands in the last decade.
(v) Relevant NGO publications such as the National Conservation Foundation and the Australian Ecological Society over the ten years 1981–1990.

From these data patterns of technological innovation or technological lineages were discerned. We devoted much discussion in our overall research design to the question of whether to return to the Aboriginal understandings that preceded European pastoralism. Whilst acknowledging the validity of this understanding we had to conclude that it was beyond our resources and brief.[7]

The second chapter (Chapter 5) in this part is based on a qualitative research study of the organisations involved in rangeland R&D at the time of our study. The main organisations have been listed in point (iii) above. As we have previously outlined, it was the naming of the problem as the failure of technology adoption and the professional and organisational concern with this perceived dilemma that had brought us to the rangelands in the first place. This concern was initially expressed as:

a limited understanding of the communication processes and resources used by sheep graziers to secure and evaluate information and the factors which influence their choice of rangeland management strategies and production technologies.

[7] Those interested in pursuing this might read Flannery, Tim (1994) *The Future Eaters. An Ecological History of the Australasian Lands and People.* Reed Books: Chatswood.

From this statement of the problem situation it was concluded that there was a need to ensure that technology transfer processes reflected the needs and limitations of these important agro-ecological environments. In the resultant formulation of our project a senior member of staff from NSW Agriculture was seconded to our team. It was from his concern about the future role of organisations like his own and the emerging theoretically informed view within the group that the organisations may have been part, or indeed more of, the 'problem' that gave rise to this research.

The data for the chapter were derived from an initial workshop conducted with twenty-four rangeland R&D staff from all the relevant institutions and a series of twenty-five semi-structured interviews conducted with rangeland R&D staff ranging from senior management to front line rangeland research and extension staff. The interviews were conducted to explore their interpretation of their R&D experiences. Where relevant, themes which arose from the interviews were further ellucidated with data from a later co-researching project with rangeland advisory staff of NSW Agriculture. Evidence had emerged from interviews with graziers and the initial workshop that the only tradition apparent amongst R&D staff was the first-order tradition. All interviews were taped, transcribed and themes identified based on grounded theory (Glaser and Strauss, 1967).

Whilst this chapter is grounded in a particular context and a particular set of institutions, it has become apparent to us that this is not an isolated story. The gap between authorities and people that is built into the structure of British public administration is one of the major conclusions reached by Susan Wright (1992) in her exploration of whether participatory development is possible in rural Britain. It is also a situation we have encountered in subsequent research in Britain, South Africa and Australia in both rural development and community-based environmental decision making and is a feature of research published by Leach, Mearns and Scoones (e.g. 1997) based on their Indian, South African and Ghanian experiences.

References

Glaser, B. and Strauss, A. (1967). *The Discovery of Theory: Strategies for Qualitative Research*. Aldine Publishers, Chicago.

Leach, M., Mearns, R. and Scoones, I. (1997). Environmental entitlements: a framework for understanding the institutional dynamics of environmental change. *Insitute of Development Studies Discussion Paper* 359, 39 pp.

Wright, S. (1992). Rural community development: what sort of social change? *Journal of Rural Studies*, **8**, 15-28.

4 From theodolite to satellite: land, technology and power in the Western Division of NSW

Adrian Mackenzie

4.1 'The facts' and the 'best possible footing'

The 'fact' is, we are told, that the semi-arid rangelands of the Western Division of NSW are no longer as productive as they were. Decades of mismanagement and overstepping limits show that, despite the best advice, individual graziers act unsustainably. Therefore present patterns of land tenure must be readjusted to fit in with the ecological limits of production. Gripped in a vice whose slowly closing jaws apply the pressure of global markets to the productive limits of the drought-prone land itself, the only solution available on these facts is to humanely extract – that is, move from the Western Division – those caught in between. On the basis of these impregnable 'facts', certain current proposals see readjustments of landholding arrangements – a kind of 'de-stocking' of grazier numbers – as inevitable.

Given the gravity of the changes proposed, there is good reason to ask how the 'facts' which support the proposals came to be. How does a social fact come to be? If it touches on the truth of a situation, it is because an arrangement of power, an apparatus set up to do things, to cause people to act in certain ways and to produce certain kinds of knowledge, makes it into the truth of that situation. Facts or truths are produced as effects of power. An arrangement or relation of power brings together different people, institutions, things, and ways of acting. Often stories (or 'narratives'), even knowledge itself – scientific, commonsense, traditional, unspoken or set out in numbers and maps – will be part of the arrangement of power that sets up the channels of action and produces the truth of a situation. Setting out to explain how a certain story – like 'the semi-arid grazing lands have been mismanaged through overstocking' – comes to be taken as a fact does not mean claiming that the fact is false, and that somewhere else we can find the truth of the situation. It means showing, as cautiously and meticulously as possible, under what conditions a particular story becomes the material and seemingly indisputable truth of a situation.

This means a change in perspective. Telling a story about how facts are produced is a different kind of challenge to the facts than the kind usually heard in debates about how the markets and the Western Division can be reconciled into productive harmony. It's a story of how certain ways of speaking called 'fact' or 'truth' only make sense within a certain organisation of action. Let's begin with the two fundamental components of action in the Western Division. Obviously there are 'people' (graziers, families, managers, rangers, ministers, officers, bureaucrats, scientists, shearers, directors, inspectors, tourists, etc.) who affect each other (e.g. persuade, suggest, advise, report, order, argue, protest and so on) through language, actions, and gestures. The general term for the ways people affect one another is 'power'. The fact that they need to persuade, order, advise or argue with each other shows, as in any social situation, there is no simple agreement of interests, or that interests will always remain the same.

Inevitably this conflict involves the other fundamental component of action – things: stock, houses, wildlife, plants, pastures, artesian bores, roads, books, money, reports, telephones, motors, acts of parliament, pipes, satellites, institutions, companies, banks, leasehold agreements, computers, databases, corporations, markets, and above all, the land itself. While it's hard to summarise all the ways in which people come into conflict, it would be accurate enough to say that for a while now, the most contested thing, the thing around which lots of other things and people are magnetised, is the land(s) of the Western Division itself. On the regulatory side, satellites, computers, databases, surveys, investigations, reports and analyses; on the grazier side, fences, pipes, bores, ploughs, fires, animals, stock, all focusing on this single complicated 'thing', the land. To tell a story about things means to tell a story of the relations of power and the ways in which people affect each other. In other words, people don't just argue about things. Things are part of the relations of power.

4.2 Lineages of people and things

What follows is a story of relations between some particular people and things in the Western Division of New South Wales – the rangelands, maps, legislation, regulations, government commissions, stock, scientific research. Perhaps it cuts out some things of equal significance in the lives of people in the Western Division – the international markets, the weather. But we are interested in seeing how the particular 'fact' of the Western Division's mismanaged lands came into being. As I said above, not simply the 'fact' of what happened in the Western Division over the last 150 years, but the relations between people, the relations of power, that produced this fact as the contemporary 'truth' of the Western Division and its occupants. Like people, things have a lineage, or a history of descent. No person, tool, machine, book, institution, terrain or ecosystem exists in a current form without descending from something earlier. Like living hereditary lineages, technologies descend and propagate through alliances with other things, by bringing in new forces and assembling new arrangements. Looking at the lineage of things can help tell a story of how the present came to be. Because they embody ways of acting and knowing, they can help us understand that the present is not a rock-solid fate, but a kind of contest which stages or operates certain relations between people and necessarily pushes other possible relations to the side. Material things such as technologies embody social relations and give them an aura of natural inevitability which is hard to shake off.

Clearly, histories and stories aren't outside relations of power, for through them, people also affect each other. They participate directly in the contemporary relations of power by saying what happened in the past. Different people tell different stories about themselves and their things. Some of them aren't widely heard (e.g. the stories graziers tell about themselves and their lands), while others are written down and (sometimes) read. One place where stories can be found about certain things in the Western Division is in the reports and publications that people in commissions, consultancies, government departments and research divisions often write. In the stacks of reports, journals and papers that line up on the shelves of archives and libraries, there are written traces of the generations, mutations and alliances between lineages of things. (No doubt they can be found in other places too: on the land, in the lives of those who dealt with them, but there the stories are harder to decipher.) A history can be written about how these stories came to be written, and about how these stories came to represent the relations between people and things in terms of the 'facts' of the Western Division we often hear today.

The assumption that I am making is that the situation or 'fact' of the Western Division today is the product of relations of power between government or regulatory bodies and graziers. We might understand this situation by concentrating on the two major lineages of things that have carried relations between people who work in government (or its appendages) and those who work on the land. The two lineages of things that have always been present in the (white) history of the Western Division are:

- technologies of survey and surveillance.
- technologies of production.

A simple justification of the selection of these two lineages can be given. From the outset, the colony of New South Wales faced two problems: one of establishing order (amongst the convicts, over the Aboriginal tribes, in an alien landscape); and one of production for the purposes of feeding the colony, of making a viable, productive independent entity. (In a sense, government objectives in the Western Division today seek to do no more than this.) The disciplinary function of the colony which I am calling 'technologies of survey and surveillance' affects pastoralism directly in ways that we will discuss below. On the other hand, 'technologies of production' (of food and commodities) were for a long time outside the domain of government intervention and only in this century gradually became a legitimate object of governmental power.

4.3 From scrub to screen: technologies of survey and surveillance

How does the set of ordering relations carried by technologies of survey-surveillance grapple with regions remote from direct government intervention? The earliest organised scientific research endeavour to reach the current Western Division was the government survey. Heavily promoted by the colonial administration, it brought the technologies of map and theodolite to bear on the imagined but previously unmanageable interior expanses. Map and theodolite were used by the surveying expeditions in order to produce directly useful scientific knowledge of the colony's geography. Useful for what? In hindsight we can say for everything from dingo control, to closer settlement policies, from taxation to woody scrub control, and now, to 'sparsification policies' and the conservation of biodiversity.

The different exploratory incursions into the interior are usually regarded as homogeneous in character. The achievements of the early colonial explorer-pastoralists – the two functions are often combined – and government survey expeditions are often confused, as if they all had the same interests. They all end up being read as part of the same process of 'opening the country up', or they're presented as a continuous progression from hazy first impressions of the explorer-settlers to the orderly scientific mapping by government surveyors of the territory. Just from reading the journals that have been preserved of those involved, we can view the itineraries of an explorer such as Charles Sturt, whose parties made the earliest forays into the regions now known as the Western Division, as profoundly different to those of a surveyor-explorer such as Thomas Mitchell, who travelled through the same territory soon afterwards. As kinds of knowing and acting, one was interested in movement, the other in possessing and controlling. The equipping and procedures of the government survey illustrates the difference: Mitchell proceeded laboriously, travelling by heavy bullock dray with flocks of sheep, carefully plotting and making inventory of all significant features by means of trigonometric survey. His methodical advance carried the full weight of an observing, measuring and ordering force (Carter, 1987, pp. 99–135).

The government funding of Mitchell's expeditions relied on the implicit promise that surveying-exploration was the first step in transforming territory into a stable and well-ordered space of settlement. Rather than discovering something new, the surveying expeditions initiated the process of re-inscribing the shape and texture of the explorer's experience in such a way as to invite and permit settlement by a sedentary population. The survey, especially the early government surveys (before the 1820s) carried out using the simple method of a rectilinear grid system, presented uniformly inhabitable space on the map and allowed the impetus of settlement and land taxation to be maintained. Even if the order and familiarity promised by a neat survey map should turn out to be a little disappointing in reality, it served the purpose of bringing settlers into the previously 'unoccupied' space, where, under the sheer pressure of circumstances, they could do little more than make the best of what was available.

So we can see that this particular lineage, the maps made by government survey, not only brought science – Mitchell, for instance, implemented the trigonometric survey – to what is now called the Western Division; it brought the Western Division into being. The act of naming an area 'the Western Division' in 1884 and producing a kind of thing – a map – not only represented order, limits, territories, but channelled people into occupying the lands they saw so neatly laid out on the flat surface of a map. Of course the explorer-pioneers still had to transform a set of symbolic lines found on the map into a life by carving out their own sense of place, their own boundaries and enclosures in the terrain they inhabited. The following events show how important the ordering role of the survey was: in an 1837 letter to Governor Bourke, wealthy established pastoralists, to whom colonial governments always turned a receptive ear, express a sense of outrage at the lack of regulation of 'the interior':

The Interior of the Colony is infested by gangs of cattle-stealers and other disorderly persons, whose depredations are carried on to an alarming extent. . . . The nefarious practices of these men are greatly facilitated by the system of taking unauthorised possession of Crown Land, or Squatting, which now prevails. . . . Many of these men are known to possess large herds of cattle which cannot be detected or prevented, so long as they are permitted to move from one part of the Country to another, and take unbounded possession of remote and unfrequented tracts of Grazing Ground.

(Letter to Governor Bourke, 1824–1837, p. 489.)

The protest seems to centre not so much on the alleged cattle stealing perpetrated by the inhabitants of the lawless interior, but on the threat of a disorderly occupation by the squatters which would leave the interior as wild and alien as it was prior to occupation. From the remoteness of a Westminster office, the Colonial Secretary, sensitive to the economic value of the pastoral activities in the colonies, responds not by agreeing to police the interior but by promising to *survey* it properly so that an orderly occupation can take place:

His Excellency will be ready to promulgate an arrangement for authorising a temporary occupation by persons from whom no aggression on the property of others can be apprehended. His Excellency is satisfied that a permissive occupancy, thus guarded against abuse, is required by the best interests of the Colony. Tracing the present unexampled prosperity of New South Wales chiefly to the production of fine wool, His Excellency is desirous of avoiding any unnecessary limitation of pasturage.

(Secretary of State for the Colonies, 1824–1837, p. 490)

From the viewpoint of State administration, only the survey can properly determine where the 'property of others' begins and ends. Maps, although they weren't the only form of exploration undertaken, are objects that generate relations between government and squatter-graziers. They allow one group of people, one set of interests – specifically the survey officers, but beyond them, colonial administrations and the imperial governments who listened to established pastoralists – to direct the movements of other people.

As well as providing knowledge of geography and topography, the map had a basic role to play in ongoing good government. The power of the survey was not just concerned with opening up the interior to colonisation, but ensuring that colonisation and settlement, wherever it occurred, would conform to the topography of good government as well as the topography of the land. Good government entails property or tenancy arrangements. Taxation or rents can be drawn only where clear property arrangements are in place. Thus a few years later (1840), the Colonial Secretary again intervenes:

The rapid extension of Settlement over the surface of New Holland renders it natural to expect that new arrangements should be necessary for the administration of its affairs. . . . One uniform price for all Country Lands renders it probable that the best Lands will be taken up first, instead of, by a difference of cost, tempting persons to begin with Lands of secondary qualities. Thus none are forced into premature cultivation, but the different Lands of the Colony are successively occupied in the natural order of their advantages. . . .

The Lands which remain in the hands of the Government in a Colony, in the circumstances of New South Wales, may be expected to be the best and the worst: the worst for obvious reasons; the best on account of having been reserved because of their special advantages.

(Secretary of State for the Colonies, 1840, p. 1–2)

Such arrangements, brought to bear via the technology of the survey, grapple the figure of the squatter, pastoralist or grazier to central government. The map as a technology for the representation of geographical space is thus allied to the State as an apparatus for the collection of taxes and revenues.

We need only look at the succession of Land Acts between 1860 and 1900, at the establishment of the Department of Crown Lands with its administrative apparatus, district surveyors and survey offices, to see how comprehensive the relations of power made possible by the survey map become. Maps are augmented by and correlated with other administrative procedures, with various leasing arrangements designed to maintain or increase the population of the Western Division. (There has always been concern, as there still is today, that too few inhabitants in the area might lead to its collapse as a governable entity.) Maps generated by the technology of the survey are linked to the records of stock numbers, rainfall, commodity output, and population distribution. Seen as things that couple people into relations of power, the survey maps and all their subsequent appendages do not simply describe a pre-given object. In a certain sense, the map makes the territory in which relations of power between graziers and government will evolve. It begins the process whereby 'being a grazier' implies certain kinds of relation to regulatory authority.

Refinement of the techniques and technology of the survey over the course of colonisation permitted a more detailed grasp of the space of the interior. During the last half of last century, the establishment of regional survey offices extended the centralised power of the survey so that remote regions such as the Western Division could be effectively settled under leasehold arrangements. The surveyors' responsibilities were amplified to include reports on the condition of the tenanted lands. The evolution of the technological lineage of the survey in the Western Division thus allowed new interests and responsibilities to be attached to the figure of the grazier. The regulatory bodies began to do more than survey and apportion land. They maintained records and archives which generate an objective context to guide action in the Western Division, so that now the actions of those who live there can be measured against a norm backed up by evidence compiled from the surveys and records.

The relations of power that invest maps have mutated a number of times – from facilitating orderly settlement, to maintaining taxability, to regulating closer settlement, and finally to monitoring land condition and ensuring sustainable landuse. The current generation of the lineage of survey-surveillance produces periodic or continual re-mappings of territory. The 'objective' representation of the territory is continually updated. Satellite imaging systems allow maps to represent current or recent changes in states of vegetation, land use and water movement.

Satellite Imagery
Within the Western Division, the Western Division Monitor continues to provide satellite imagery routinely over the areas most likely to be developed for clearing or cultivation. Images are perused for anomalies which are then investigated by field officers. Few anomalies prove to be illegal clearing or cultivation but the process allows the Commission to demonstrate sound stewardship of the land resource to the community.
(Department of Lands, 1989–90, p. 63)

The contemporary lineage of the map, more likely displayed on a computer monitor than on paper, connects ideas of 'land management', couched in the terms of ecological discourse, with the grazier as 'husbander' of an ecosystem or a fragile resource. Current surveys operating through remote sensing systems, through management surveys such as the Soil Conservation Service's Property Resource Plans, and through the various rangelands monitoring programmes promoted by the CSIRO and other government agencies still map space in order to regulate the 'proper' occupation of territory. The only changes concern the definition of 'proper' occupation. The shift has been from a desire for orderly settlement and taxation to orderly management of the rangelands as an ecosystem. The technology of the survey, albeit effected through LANDSAT data, spectral reflectance and computer image analysis, and manipulated in computer databases rather than on charts and ledgers, still focuses on a refined and increasingly quantifiable space. Despite the massive changes (which we will have to account for), there is continuity in the technological lineage of the survey: whether carried out through theodolite or satellite, it consistently links the representations of land to the actions of graziers.

The position of the grazier as 'manager' and the representation of territory as 'ecosystem' departed long ago from the specific intentions of the initial government surveys, which centred on defining a space of taxable and productive occupation; long ago the state-defined responsibilities of the grazier and the nature of the survey turned towards creating a productive manager and lately, towards creating an ecologically responsible one. (Later we will see how the emphasis came to shift from creating a taxable space of occupation to the current preoccupation with managing sustainable ecosystems.) Throughout the evolution of this technological lineage of the survey, the scope and limits of individual modes of action, and particularly those of the lessees of Crown Lands, have always been at stake. Not only have relations of power sought to limit or control, they sought to produce particular ways of acting that we now designate with the term 'grazier'. Seen in terms of relations of power, the meticulous preparation of maps over almost one and half centuries has not only opened channels of observation and monitoring between government and its territory, they have been used to arrange and produce occupancy of the Western Division itself. Survey and monitoring efforts have never been distinct from the regulation or control of how spaces such as the Western Division are occupied or inhabited. They not only lay out the groundplan of settlement (the drawing-up of leases and boundaries requires maps), they later function as the means of monitoring more closely how land is being used or affected by people like graziers.

4.4 The land itself

We have seen that from the earliest days of colonial settlement, control of population and property had been the main *administrative* problem. Aside from the lawless squatters mentioned in the 1837 letter to Governor Bourke, the annual reports of the Chief Stock Inspector to Parliament in the second half of the nineteenth century often complain of the failure of graziers to send in their annual stock returns:

I have the honour to submit herewith my Report for 1874, and in doing so I would once more call attention to the very imperfect manner in which the information asked from the owners of stock has been given. In a great many instances the forms distributed were not returned, and in others the answers they contained were of the most meagre description.

(Chief Inspector of Livestock, 1875, p. 717)

While the Chief Inspector obviously had other duties concerned with disease control, his relation to the graziers is vexed by administrative obstacles.

At the end of the nineteenth century, a new concern surfaces in government reports. The famous and still-quoted Royal Commission of 1900–1 shows a shift in emphasis from people or, strictly speaking, lessees, to the lands of the Western Division itself. The lands of the Western Division become a problem in their own right, a problem distinct from other regions of agricultural production. We might say that the institution at this time of the Western *Lands* Board on the basis of the Royal Commission's recommendations symbolises the change in emphasis. In the new epoch of power relations, a new and distinct object of regulatory interest comes to the fore: the *productivity* of the land itself. An elaborate cavalcade of technological lineages and administrative bodies concerned with the specific limits on production in the Western Division starts to arrive. What had previously been simply one amongst many districts administered by the Department of

Lands and Department of Agriculture emerges as an entity, characterised by special problems of productivity.

This unique new entity, the Western Division, will come to be represented and regulated in the interests of production, rather than as an area difficult to tax and survey because of remote vastness. The shift represents a mutation in the relations of power, and correspondingly, a re-definition of what it is to be a grazier. No doubt, producing at least a livelihood for themselves and for the family communities they lived in had always been a central preoccupation for graziers. Of necessity, they had been active in adopting, borrowing, applying or inventing mechanisms, breeds, motors and materials to the process of producing a livelihood. Crucial transformers of pastoral action such as the steam engine, arterial bore, windmill, fencing wire and iron water pipe, entered the Western Division without any technological drive on the part of State-sponsored research and regulation (see Chapter 3).

The references to a number of other grazier innovations which filter through government reports can be seen in the same context:

- The gradual increase in the number of watering-points per property beginning in the last century directly reflects this changed mode of action. The modes of action appropriate to being a grazier no longer relied so much on influencing the actions of others (shepherds), but in controlling stock directly.
- In conjunction with galvanised water piping, the wire fence, a somewhat later innovation in boundary technology, was used and promoted by some pastoralists as a means of effecting different distributions of stock across the rangelands, and in particular, diminishing stock concentrations around central watering points (see Lange *et. al.*, 1984, pp. 46–54). The terms in which these pastoralists viewed boundary technologies, particularly in South Australia where more flexible leasehold conditions applied, anticipates the organised research efforts begun in the last twenty-five years to ascertain the role of fenced boundaries in grazing management and to specify guidelines for their deployment. (What grazier experience had already at least begun to accept in the use of boundary technologies nearly a century ago is now being re-formulated in the parameters of computer simulations of the effects of fences on grazed lands.)

In these few examples, crucial innovations such as fences, watering-points and water-piping appear as things which change the movements of people and animals, both introduced and native. Fences make shepherds redundant, watering-points and water-pipes mean that stock, feral animals, graziers and wildlife alter their patterns of movement.

Hence there is no doubt that prior to the time when relations of power begin to concentrate around the problem of the productivity of the land itself, graziers were not passive victims of circumstance, hidebound traditionalists but active innovators. Indeed technological innovations introduced by graziers are sometimes later taken up by regulatory bodies and are overwritten by new relations of power. We can observe this process in the case of the dingo fence: the first extensive border fences against dingoes from south-west Queensland were the result of the combined efforts of graziers themselves. Only later, does the body now known as the Wild Dog Protection Board take over the task. The evidence is also convincing when we turn to measures related to productivity of the lands. In the 1870s, some pastoralists experimented with introduced grasses ([Chief Inspector of Livestock, 1875:721] mentions introduced grasses). Presumably these trials, carried out well before drought and rabbit infestation during the years 1890–1900 brought the productivity of the land itself into the spotlight, indicate some grazier attempt to regulate the productivity of the land. Later mentions of graziers trying to cultivate pastures with introduced grasses appear in 1959 and 1960:

Increasing numbers of graziers are carrying out practical experiments aimed at improving the carrying capacity of their land and making it safer during the recurrent drought conditions typical of the Western Division. . . . Some enterprising settlers have established small plots of Buffel grass and by their experiments are making a valuable contribution to the search for drought resisting grasses suitable for light rainfall areas. Every encouragement is being given to these settlers.

(Western Lands Commissioner, 1959, p. 38)

We might also cite the evidence which appears as early as 1907 of grazier-initiated efforts to obtain a virus which would ennervate rabbit reproduction. (Ironically enough, the immediate reason for the failure of these experiments seems to have been New South Wales government bans on inoculations being tested on the mainland. The Pasteur Institute scientist sponsored by the pastoralists returned to France, leaving the rabbit viruses untested: (see Select Committee, 1922, pp. 220–221). Again the first reported uses of goats for shrub management, meat and fibre production credit graziers with the innovation (Western Lands Commissioner, 1972, p. 31). The fact that low-profile grazier-initiated practices or innovations often resurface much later in the domain of scientific investigation can be explained by the shift in relations of power. When the object of regulatory interest shifts from proper occupation to the productivity of the lands themselves, innovations such as waterpipe which once held no real interest for regulatory bodies, become a factor whose precise effects must be quantified and modelled.

This is not to say that the State had not been concerned all along with the task of ensuring that productive lands were linked to markets: it maintained stock-routes and wells, as well as laying rail to certain points precisely in order to allow stock and wool to reach markets. But these measures, as extensive and more or less useful as they were, never referred to or addressed the productivity of the lands of the Western Division themselves. The fact that they were State-owned and run shows that they were conceived as part of the administrative apparatus of the State. It follows that when, in the course of droughts, the stock-routes were over-taxed by the grazing pressure of itinerant mobs of stock, the problem was seen as one of improper use of stock-routes (i.e. an administrative problem), not one of inherent limitations on the productivity of grazing lands in the Western Division.

All of the examples of technical or technological innovation by non-scientists stand apart from the organised investigations which state-sponsored research began to carry out when the land itself became a problem. They stand apart not because they are directed at a different object. They too are directed at affecting or transforming relations between stock, pasture, wildlife, and feral animals towards the side of production. They too involve representations, habits and actions in relation to the land. But they are separate from the administrative apparatus. Indeed they and many others do not appear in the reports, papers and records, or in the whole machinery of knowledge administration which slowly moves into action as every aspect of the land, plant and animal life is opened to scientific investigation.

When productivity becomes an object of power in its own right, government regulations, research organisations, extension officers and so on begin to participate in the definition of the methods, materials and limits of production. Science, for the first time, has open to it a whole panoply of objects aside from the mostly static objects of cartography and survey. It no longer need simply chart the surface of the land and its contours, it could start to measure, analyse, categorise and render calculable everything from the subsoil up. The interaction between vegetation and animal life in particular occupy the prime focus in the new optics of research. What we would like to see is how the scientific representation of these objects tends to overlay and re-contextualise the expertise and practices that had previously been the domain of tradition and local knowledge.

We can observe the crucial shift in the object of power more closely by discerning two broad lines of descent in the technological lineage of production. If there is a split between the two branches of the lineage, it is clearly derived still from the division in regulatory interests. Thus the Soil Conservation Service and the Department of Agriculture were seen as having, in principle, different objects, and therefore deploying different technologies. On the one hand, technological lineages of the pasture act on the surface of the land itself, its vegetal coverings, its rupturing, blistering, and scalding by water and wind. On the other hand, the lineages of animal life act on life-forms that move across and beneath the surface, dynamically acting upon it through foraging and territorial patterns. These two objects of the technological lineage of production – pasture and animal – have been in dynamic interplay since the shift to the new object of power relations at the beginning of this century. As we will see, only later do ecological understandings bring the two branches of the lineage back together.

4.5 Technologies of animal life

If technologies of animal life act directly on the bodies of animals, it is generally to either 'improve' reproductive and productive growth in the case of stock, or to hinder reproduction and growth in the case of parasites, feral and competing wildlife. This technological lineage targets both productive and non-productive (i.e. competitive or parasitic) life-forms. Animal technologies grounded in science are usually not specific to the context of the Western Division. Understandably, they import a more general understanding drawn from biological sciences of animals as living systems susceptible to technological intervention and modification. The plethora of investigations around the physiology of wool production, the nutritional and dietary regimes of various native and introduced animals (see Fowler's Gap Field Station, 1969, p. 14 for an example of the kind of work done on the dietary regimes of kangaroos using plasma and urine electrolyte samples), population studies of wildlife and feral animals, studies into genetic modification of sheep blowfly (see for instance Wool Research Trust Fund, 1991, p. 144), enhancing sheep immunological defences against blowfly strike, immunosterilisation of rabbits (see Division of Wildlife & Ecology, CSIRO, 1988–90, p. 8), and so on, all have in mind the same possibilities of prediction, manipulation and control of living systems.

These 'leading-edge' technologies represent animals – and fibre production – as susceptible to interventions which induce changes in or directly modify the living structure. Genetically engineered and immunological techniques are the most clear examples of these developments. In this respect, technological innovation in the Western Division parallels the trend evident in many spheres of research and production, agricultural or industrial. But there is a significant difference. Technologies introduced into the Western Division by regulatory or research bodies are, almost without exception, framed by a context in which land management is administered by the authorities, who base their authority to act upon the actions of others on their privileged access to techno-scientific expertise.

Technologies which manipulate the structural character of the animal life involved in pastoral production call for meticulous technical controls and monitoring. Their efficacy may sometimes be immediately obvious in use, but the possibilities for the grazier to decide how such technologies as an immuno-assay nutrient kit or a genetically modified blowfly will be used are, to say the least, limited. The scope of a meaningful grazier response if the technology breaks down are also narrowly defined. These technologies position the grazier in the process of wool production as a worker, a manipulator of pre-designed tools controlled from elsewhere. Graziers in general could not be expected to be in a position to deal in detail with the techno-scientific attributes of the technology they are encouraged to use. Improving productivity has the effect of transferring the interplay between the graziers' perceptions and the institutionally embedded techno-scientific discourses significantly further towards the latter pole.

Constraining action to occur according to scientific parameters motivates another scientifically grounded technological lineage. The last decade or so has seen the development of various expert systems or management packages such as RANGEPACK, SHRUBKILL or GrazFeed, which directly integrate the results of nutritional studies of livestock, ecological studies of vegetation systems and cost-benefit analysis techniques in 'decision support' packages intended for the use of graziers and land administrators in 'coping' with the wide array of variables pertinent to rangelands production. Expert systems and decision support systems are perceived as a means of directly 'transferring' a set of management practices, based on sound scientific data, across the gap between the institution and the grazier. The principal goal of these systems is clear: to enmesh the range of actions available to the wool-grower with the programme of research carried out by the regulatory bodies.

Aim
To develop management packages that can be used by graziers to judge the likely impact of strategic and tactical decisions on farm productivity and profits. The packages offer graziers a consistent way to check their own intuitive assessments of a new farming technology.

GrazFeed was released . . . in July 1990. It is already in use in 65 centres throughout Australia by agricultural consultants, extension officers, veterinarians, graziers and educational institutions.

(Wool Research Trust Fund, 1991, p. 211)

From our perspective – an interest in how technologies participate in the fabrication and maintenance of relations of power which produce an object of management (the rangelands) and a manager-subject (the grazier) – the 'success' of these innovations signals not so much that the technology of computer modelled management is well-designed, but that the relation between regulatory interests and grazier interests has altered its character. The fact that in the reports of success graziers are only one amongst the many institutional users itself indicates the variety of regulatory interests focused on grazing.

4.6 Regulating the pastures

As institutionally driven interventions, the technologies of the pasture focus on the general space of production rather than directly on the animal life in or on it. Their purpose lies in 'pasturing' the lands of the Western Division so that they can be brought into production. By altering the distribution of edible vegetation, by diverting waterflows, by re-shaping the very contours of the land itself, they mean to ensure that productive animals can be buffered against non-productive competitors, protected from seasonal extremes and given greatest possible access to edible vegetation. In their affinity to technologies of animal life, the technologies of the pasture sought to increase productivity.

In order that the land itself – as opposed to property arrangements, taxation or population goals – could initially become an interest or object in its own right, new ways of representing, measuring, analysing, and speaking about land were needed. The nature of this shift can be seen if we compare the language of the Royal Commission of 1901 and the language of a contemporary scientific report. The Royal Commissioners write:

1. The Causes of the present depression in the western division.

(a) The low rainfall and the frequent subjection of the country to periods of drought may fairly be regarded as the primary and most constant causes of the difficulties which beset the western grazier. (Royal Commission, 1901, p. 136)

'We regard the affairs of the Western Division as in a very critical condition, and think that, in order to avert disaster and put things on the best possible basis, a Board consisting of, say, three men, of the highest procurable ability, should be charged specially with the duty of dealing with the problems

which the Western Division presents, and the task of re-adjusting existing conditions.'

(Royal Commission, 1901, p. 154)

Drought, rabbits and overstocking were the salient features by which the lands of the Western Division were characterised in 1900. At that time 'three men, of the highest procurable ability' were to deal with the dire threats to productivity (i.e. the output of wool) posed by these elements. Compare a typical description of the problems of the Western Division today:

Grazing management for the long term stable production from the rangelands, and particularly in relation to desirable pasture species, is not well understood and should be the target for future investigations into the area of rangeland restoration and management.

(Green, 1989, p. 115)

Here the general problem is one of understanding how the rangelands should be managed and restored. In this context, the problem becomes one of understanding 'natural limits' of the rangelands themselves rather than directly acting to alleviate existing problems which restrict production. The limits on production imposed by climate and vegetation are no longer something to be circumvented, but become a problem of knowledge and an object of contestation in their own right. No longer are technologies which operate to increase yields – such as the technologies of the animal lineage – immediately acceptable. The precarious nature of the limits on production themselves must first be understood. As an objective of technological intervention, increased production is for the most part secondary.

How did the shift in regulatory perspective from increasing productivity at the turn of the century to the contemporary problems of – in effect – discovering the limits of production come about? Scientific research undertaken in universities and government research bodies is the most energetic influence. The capacity to generate an understanding of the natural limits on grazing is primarily the province of those who can run trials to establish limits, invest in unproductive experiments and communicate the results in a shared language and format (the scientific report or paper). The responsibility for defining problems in relation to the productivity of the land has increasingly come to rest with scientists and the extension officers conversant in scientific methodologies. Graziers may have made the bulk of the contributions to the prevalent understanding of the productivity problems of the Western Division in 1900, but within forty years the sound of their voices barely murmurs behind the reports of the new generations of research and regulatory bodies. When they are heard, their actions seem curiously reminiscent of other interests.

On the face of it, the following report sounds encouraging:

Graziers around Broken Hill have investigated pitting, furrowing and banking as reclamation techniques to increase both the stability of their degraded country and the availability of drought feed. The Service has assisted by establishing the Broken Hill Land Reclamation Demonstration Program.

(Soil Conservation Service of NSW, 1986, p. 40)

But this 'initiative' on the part of the graziers doesn't mention the fact that from 1940 onwards, the Soil Conservation has been propounding just these kinds of measures to graziers as the best way to manage their erosion troubles. From the beginning of its efforts at controlling soil erosion through earthworks and checkerboard ploughing, the Soil Conservation Service of NSW, one of the first research and regulatory bodies directly active in land management in the Western Division, developed technologies of pasture control and regeneration which largely relied on the grazier accepting the expert advice of Service officers and allowing the Service to implement its own recommendations by bringing in earth-moving machinery. Even in the previous example of pitting, furrowing and banking for erosion control 'investigated' by graziers, the Soil Conservation Service reports that:

The investigation and demonstration program will be undertaken over a four year period, and will develop technology, demonstrate to landholders the effectiveness of furrowing as a reclamation measure and make available the resources necessary for widespread implementation of furrowing programs.

(Soil Conservation Service of NSW, 1986, p. 40)

Related technologies, such as water-ponding and contour furrowing, still rely on graziers participating in the Service's reclamation programme rather than implementing the technology themselves. Service officers, for instance, carry out the laser-aligned levelling required for water-ponded regeneration of pasture (Soil Conservation Service of NSW, 1986, p. 41) and resources are 'made available' for the implementation of furrowing programmes. While ultimately graziers bear a large financial responsibility, the implementation of these technologies depends on the presence of institutional arrangements which provide for their development and promotion, and in doing so, confirm the 'rangelands' as a problem which necessitates institutional intervention. The mechanical land reclamation techniques promoted by the Soil Conservation Service over the last 50 years in the Western Division consistently position the grazier as needing assistance in dealing with the fragile land resource and in occupying lands degraded through past grazing practices. These technological interventions are undeniably successful in themselves. But the form of these successes – land reclamation techniques, fire regimes, advice on rangelands management – ensures that the regulatory body is also in a position to monitor the space in which its technologies are applied.

In this sense, technologies of the pasture necessarily import relations of power into the arena of production. The form of these relations are generally the same regardless of the technology, technique or plan involved. Since expert or scientific knowledge comprehensively describes the situation in advance of any grazier participation, the technologies developed by institutional research and development for managing problems of the grazing lands do not generally depend on any knowledge or skill on the part of the graziers themselves. Rather than depending on grazier action, the knowledges and technologies developed in response to the problems of the land itself install a reliance on the monitoring of the regulatory bodies concerned. The predominant outcome when institutional arrangements are designed to 'prevent over-use of the land which may lead to pasture degeneration' [Department of Lands, 1988, p. 50] is that the monitoring functions of the regulatory bodies are correspondingly enhanced in importance. The 'problems' of the Western Division are thereafter seen as manageable only through the establishment of a centralised information base concerning all 'relevant' aspects of the land resource:

Clearly, information and quantitative data of the human environment and natural resources are required for the effective planning and management of the Division.

Thus, we firmly believe that a sound information base concerning the resources of the Division is essential for proper development and management, and as a basis for determining future changes in land condition.

(Joint Select Committee, 1983–84, p. 34)

As this statement makes clear, 'relevant information' for management purposes basically means quantitative measures, information susceptible to statistical treatment. Each further step in the scientific understanding of the problems of the rangelands opens the possibility of a new mode of institutional action to alleviate that problem and thereby weaves a new thread into the net of relations that enmeshes grazier actions with the monitoring functions of the institution.

In re-defining the problem of the Western Division in terms of the land itself, rather than the people or population that inhabits it, the lineage of production technologies necessarily acts not only on the land itself. Sometimes, the grazier comes to be seen as a 'constraint' to land management, an actor whose actions call for objective and independent monitoring as another factor in the rangelands ecosystem. Here monitoring is implemented in order to circumvent or supplant grazier experience with an independent channel of observation:

A monitoring system is an integral part of the management process. Without such a system it is not possible to determine if management strategies are fulfilling their intended objective. At present both conservationists and administrators have little factual information on the long-term change in the condition of the arid lands.

(CSIRO Rangelands Research Unit, 1981, p. 4)

The lineage of pasture technologies attempts to bring that other more unpredictable factor, people, within a relation of power that ensures that the land can be objectively understood as a resource to be managed. In a sense, defining the problem in terms of the land itself, and then setting out to describe, measure and analyse that problem in its own right, has necessitated re-defining the action of people as a kind of 'independent variable'. Their role has been re-formulated as 'managers' or lately, as 'custodians' of the lands they inhabit. In any case, the monitoring and control functions that regulatory bodies exert through their promotion of particular land management techniques should not be seen as simply counter to the interests of the graziers. In part they re-define those interests and thereby re-define what it is to be a grazier.

The constant pressure to manage land holdings according to the objectives defined by regulatory interests necessarily creates a situation in which graziers expect the institutions involved in the Western Division to develop and often even operate technologies which will assist in the realization of the land management goals, goals they have learned to acccpt not only as the legal condition of their leasehold, but as promising the possibility of productive and profitable habitation. Understandably, the emergence of this expectation on the grazier's part entrenches the reliance on institutional interventions – technological and managerial – in the Western Division. The definition of the problems of the Western Division in terms of 'land management' means that the major technological solutions promoted by bodies such as the CSIRO, Soil Conservation Service, and Western Lands Commission inevitably bring rangelands management more directly and in far greater detail than ever before into the purview of both the graziers and the land administrators. The various rangelands monitoring systems promoted by these bodies as a key element in effective land management inevitably provide assessments to both graziers *and* the regulatory bodies.

The effort to encompass all aspects of pastoral activity and the ecosystem it occupies within a detailed scientific model often threatens either to incorporate the grazier as a parameter or constraint on the management of the rangelands or to place the grazier outside the ecosystem altogether. This extreme result is most apparent in certain ecosystem models which, in their efforts to articulate management for conservation and biodiversity purposes, deal with conflicting pastoral production interests by limiting grazier action in the rangelands to the ecological limits of the natural system involved:

In the semi-arid lands of Australia an extensive area of more-or-less natural vegetation is grazed by sheep and cattle. Despite the agricultural purpose of this land use, management is limited to ecologically based manipulation of vegetation and animals. The landscape remains in a semi-natural state and management seeks to prevent undesirable changes in the soil, the plant cover and the animal population rather than directly enhancing productivity.

(CSIRO Division of Wildlife & Rangelands Research, 1984, p. 33)

Similarly, the contemporary managerial wisdom of simply reducing grazier numbers, of 'de-stocking' the human population of the Western Division as well as the stock numbers, is largely a consequence of the modelling of the grazier as a passive obstacle to ecological diversity and sustainable production. The persuasiveness of this viewpoint calls on the systems of surveillance and monitoring to provide evidence that – it is claimed – speaks for itself of the ecological damage suffered by the rangelands at the hands of graziers. Given its conception of the rangelands as a fragile ecosystem which can sustain only minimal intervention, it is understandable that strongly ecologically based management will turn to 'managing' those whose remaining scope of action represents more than a minimal intervention in the ecosystem they occupy. The privileged value given to preserving biodiversity does not always include the socio-diversity of human social groups.

4.7 The story's ends

Let us try to put all these developments once again into the perspective of people in relation, and of things as crucial agents in defining relations between individuals. The story of the Western Division I have told concerns a fairly narrow but complex set of things and people. The people involved are the graziers who live and work on the land and the administrators, researchers, officers and technicians of the regulatory and research bodies who work at monitoring and managing the Crown Lands. There has long been a contest over the Western Division, in which the meaning of its occupation and uses are disputed in terms of notions of proper settlement, productivity, management, and conservation. Through a complex interplay, this contest produces the mode of action which we know today as the grazier.

Technologies of power such as the survey arrived in the Western Division on the heels of the first squatters. They produced more than lines on a map. They set down the first lines of contact between State and individual graziers, and in doing so begin to define what 'being a grazier' is. From the standpoint of the State – and this is the standpoint that even the most 'purely' scientific reports nearly always take up or support, even today – the idea of the Western Division entailed the objective of orderly occupation and good government of the interior. The administrative technologies of stock reports, Land Acts, rent, taxation and leasehold agreements wove the threads of power relations into the fabric of inhabitation. To live in the Western Division – even today – is to be subject to these arrangements, although the definition of 'orderly' occupation has been re-formulated in the terms of 'good management'.

Each crisis in the history of the Western Division sets off a re-alignment of relations of power. When the very potency of the nineteenth century arrangements for producing a pastoral population triggers a catastrophic decline in productivity around the turn of the century, productivity itself arises as a new objective for State regulation. In this turn towards the system of vegetation and animal production, scientific research and management are brought to bear through new regulatory agencies and a new panoply of technological interventions. The new system of power relations used existing technologies such as the survey and tenancy agreements to effectively act in the Western Division, but on a different object of power, the land itself. This arrangement coupled the lineages of the survey with techniques developed to alter the condition of the land, its vegetation and animal life, in accordance with the goals of improved productivity. Under such arrangements, State agencies could legitimately monitor not only the the grazier, but the grazier's practical relation to the land itself. Indeed, it is perhaps true to say that power extends itself more completely through these indirect and diffuse channels concerned with managing the land than through the direct but narrow channels that originally applied to the grazier as a leaseholder and tax-payer.

It is obvious that today another discontinuity or break in the story of power relations is unfolding. Once again, a crisis brings change into the open. Where regulatory management and monitoring were once drawn on by the dreams of increasing productivity, the last twenty years have been characterised by an increasing emphasis on limits on productivity. In the context of radical limits on the capacity of the Western Division to sustain production, monitoring the state of the land and defining productive capacities on the basis of detailed ecological work has become the principal regulatory objective. Instead of increasing production, regulation seeks to manage the lands of the Western Division so that production or use of the land stays within limits prescribed by prevailing ecosystem models. In the new – but old – contest over how the Western Division is to be occupied, the stories told about land represent the grazier as another variable to be managed into conformity with the productive limits of the land. The contest remains one of seeing how the limits of the social activity of grazing can be creatively reconciled with the 'facts' that regulatory activity and research administer.

General references

Carter, P. (1987). *The Road to Botany Bay*. Faber & Faber, London.
Dreyfus, H. and Rabinow, P. eds. (1983) *Beyond Structuralism and Hermeneutics*, 2nd edn. Chicago.
Foucault, M. and Rabinow, P. eds. (1986). *The Foucault Reader*. Penguin.
Gadamer, H-G. (1982). *Reason in the Age of Science*, tr. F.G. Lawrence. MIT Press, Cambridge.
Gordon, C. ed. (1980). *Power/Knowledge: Selected Interviews and Other Writings*, Harvester Press, Sussex.

Western Division references

Chief Inspector of Livestock (1876). Report to the Minister of Lands for 1874. *NSW Votes & Proceedings 1875–1876*, vol. 6.
Department of Lands (1988). *Report of 1987–88*, Strategic Plan for the Western Division.
Division of Wildlife & Ecology CSIRO (1990). *Report of 1988–90*.
Division of Wildlife & Rangelands Research, CSIRO (1984). *Report of 1982–84*.
Fowler's Gap Field Station, University of NSW (1969). *Annual Report 1969*.
Governor Bourke (1837). 'Unauthorised Occupation of Crown Lands. *NSW Legislative Council Votes & Proceedings*, Sessions of 1824–1837, p. 490.
Green, D.R. (1989). Rangeland restoration projects in western New South Wales. *Australian Rangelands Journal*, 11(2), pp. 110–16.
Joint Select Committee Enquiry into the Western Division (1984). Third Report of the Joint Select Committee Enquiry into the Western Division. *NSW Parliamentary Papers*, 1983–84, 4th Session, vol. 3.
Lange R.T., Nicolson, A.D. and Nicolson, D.A. (1984). Vegetation management of chenopod rangelands in South Australia. *Australian Rangelands Journal*, 6(1), 46–54.
Letter to Governor Bourke (1837). Unauthorised Occupation of Crown Lands. *NSW Legislative Council Votes & Proceedings*, Sessions of 1824–1837, p. 489.
NSW Department of Lands (1990). *Annual Report*, 1989–90.
Rangelands Research Unit, CSIRO (1981). *Report of 1980–81*. Deniliquin, NSW.
Royal Commission (1901). The Condition of the Crown Tenant in the Western Division of NSW. *NSW Parliamentary Papers 1901*, vol. 4.

Secretary of State for the Colonies (1840). Despatch from the Secretary of State for the Colonies to the Governor re Division of the Territory of NSW and Sale of Crown Lands. *NSW Legislative Council Votes & Proceedings*, Session of 1840, p. 1.

Select Committee on the Conditions and Prospects of the Agricultural Industry (1922). Final Report from the Select Committee on the Conditions and Prospects of the Agricultural Industry and Methods of Improving the Same. *NSW Parliament Legislative Council.*

Soil Conservation Service of NSW (1986). *Report of the Soil Conservation Service of NSW for 1985–86.*

Western Lands Commissioner (1959). *Annual Reports of the Department of Lands.*

Western Lands Commissioner (1972–3). *Report for 1971–72, NSW Parliamentary Papers*, 3rd Session, vol. 1, p. 795.

Wool Research Trust Fund (1991). *Projects Supported During 1990–91*. Australian Wool Corporation.

5 Experience, tradition and service? Institutionalised R&D in the rangelands

Raymond L. Ison[8]

5.1 Introduction

In the NSW Western Division, as in most parts of the world, there are a range of people from different organisations who see it as there business to be involved in shaping or managing what happens to people in local contexts. One might well ask who runs the rangelands of NSW (Beck 1991)[9]? As shown in Figure 5.1 there are a range of agencies and interested parties in the NSW rangelands with varying degrees of involvement and perceived influence. NSW Agriculture, the State-funded agricultural research and extension organisation, along with the former Soil Conservation Service (SCS)[10], have been the major players with a traditional R&D presence in the rangelands. They are commonly referred to as 'service organizations' (although as our research reveals there is not always a clear answer to the question: Service for whom?).

This chapter is based on two distinct but related pieces of research. The first is a series of semi-structured interviews conducted with staff from the service organisations with a R&D presence in the Western Division of NSW (NSW Agriculture, the Soil Conservation Service (SCS) and the Western Lands Commission (WLC)). Both field staff (referred to as advisory or extension officers) and managers were interviewed. The second research project involved working with a group of rangeland R&D staff in an attempt to develop a second-order, or co-researching approach to their own work. Chronologically the latter research was initiated after our co-researching project with pastoralists (Chapters 6, 7 and 8) and is not discussed in detail here (see CARR, 1993). However, aspects of it are drawn upon where it illuminates themes which emerged from the interviews.

The interviews explored the nature of the task of 'extension' in the semi-arid rangelands as well as inviting reflections on individual's experiences of

---

8 / This chapter was written by the author based on an analysis of transcripts of interviews conducted by Peter Davey with the assistance of Lynn Webber and an initial draft prepared by Mike Hannibal.

9 / In posing this question we recognise that public sector organisations are not the only stakeholders in the rangelands – for the purposes of this chapter we choose to consider only this subset of possible stakeholders.

10 / During this research the Soil Conservation Service and the Western Lands Commission were merged into one organisation – CaLM (Conservation and Land Management). This has since been reversed.

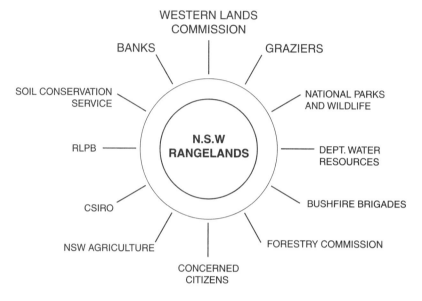

Figure 5.1
A depiction of the many agencies and interested parties that interact in the management of the NSW rangelands. RLPB, Rural Lands Protection Boards; CSIRO, Commonwealth Scientific and Industrial Research Organisation. (Source: After Beck, 1991, with permission from NSW Agriculture.)

the rangelands and their concerns and/or aspirations for the future. More importantly perhaps, these interviews explored the relationship between the ethos, structures and culture of organisations and the way in which officers felt they could do their jobs. NSW Agriculture was the organisation of focus. Many of the officers were subject to a complex set of tensions. These tensions arose from their perceptions of what was expected of them 'by head office' and the contrast between this and their own experiences and beliefs about how to do 'extension' in the rangelands. Tensions like these are widespread in organisations associated with research and rural development but have been much neglected in research conducted to achieve more effective R&D (Wright, 1992; Chambers, 1997; Blackburn & Holland, 1998; Röling and Wagemakers, 1998). As Swift (1995 p. 154) notes 'organizations and institutions are the two crucial structural components of pastoral administration'. These may be formal or informal. Together they form part of the set of constraints to meaningful human activity, which also includes environmental risk and resource availability.

5.2 Organisations, institutions and structures

Unfortunately there is much confusion in the literature and in everyday conversation about what is meant by 'organization', 'institution' and 'structure'. Swift (1995), following North (1990), describes institutions as the 'rules of the game in which individual strategies compete . . . [they] include any form of constraint that human beings devise to shape interaction[11]' (e.g. formal arrangements such as law or informal arrangements such as customary land tenure). From their perspective 'organisations are groups of individuals bound by some common purpose to achieve objectives' and 'organizations operate within the framework – the rules and constraints – provided by institutions' (e.g. a government department or a pastoral association). Swift's (*ibid*) rationale for attempting to be clear about these distinctions is so that institutional and organisational analysis and design can be conducted. His aspiration is to encourage the evolution of new organisational and institutional types which 'reduce transaction costs and enhance the benefits of particular types of action in the pursuit of pastoral development objectives'. Our aspirations in doing this research were not dissimilar.

North's (1990) distinctions between 'organisation' and 'institution' are initially helpful but from a systems perspective do not go far enough. A particular distinction which arises from systems studies is between the *organisation* and *structure* of a system. The *organisation* of a system is defined as a particular set of relationships, whether static or dynamic, between components which constitute a recognisable whole – a recognisable unity as distinguished by an observer. Organisational relationships have to be maintained to maintain the system – if these change the system either 'dies' or it becomes something else. However the *structure* of a system is defined as the set of current concrete components and relationships through which the organisation of a system is manifest in particular surroundings. Thus for a particular R&D organization (such as NSW Agriculture) the key organisational relationships might be considered to be those between politicians, researchers, administrators, extension officers and agricultural/horticultural producers (experience suggests that consumers are often excluded). If these relationships cease to exist then that which is unique to a particular organisation ceases to exist. If it were a biological organism this would mean the death of the organism but because organisations are not biological organisms those involved could choose to become some other organization remembering that the *same* organisation can

---

11 / Elsewhere the term 'social technologies' has been used to describe the same phenomenon (Open University, 1997); when particular social technologies become everyday or background and continue to be used uncritically then it could be said that they have become institutionalised – they form a tradition.

realise or manifest itself through *different* structures. Structures in this example might include particular divisions, programmes, practices (e.g. promotion practices). From our perspective it is particular structures that become institutionalised in ways that create a tradition out of which people think and act, often without reflection (see the introduction to Part I).

A trap in thinking occurs when the differences between organisation and structure are not recognised, particularly in the design of new types of organisation or the development of new structures which are capable of maintaining the existing organisation in a changing environment. This view of organisation moves beyond seeing organisations as just comprising people who share a common purpose and who behave in goal-seeking or objective-seeking ways. For the purposes of this chapter NSW Agriculture and the like will be referred to as organisations. It is recognised that what constitutes any particular organisation in terms of the relationships that are conserved is not a question that is easy to answer, and that answers are likely to differ with perspective.

In the remainder of this chapter the dominant themes emerging from the interviews are explored.[12] These themes and the implications for future organisation of R&D and the practice of 'extension' in the rangelands are then discussed in relation to the distinctions between organisation and structure.

It is inevitable that in writing a chapter such as this, the author's views may be at variance with the views of other individuals in the organisations, or that generalisations which have necessarily to be made may not represent the beliefs and behaviour of all members of these organisations. It also needs to be recognised that the data are drawn from a particular historical moment and that a narrative generated about the organisations now may be different. Nevertheless the material presented and the conclusions drawn in this chapter, are indicative of the complex dynamics which these interviews and subsequent attempts at co-research reveal within NSW Agriculture and other service organisations. Evidence provided by other researchers (see those cited above) suggest that the phenomena we encountered are widespread in rural R&D organisations.

5.3 From the 'coalface' to the executive suite: themes that emerged

Three groups of people were interviewed for this chapter: 'coalface' extension personnel, middle managers and senior executives.[13] The first two of

---

12 / Where it contributes to the argument, material from interviews has been quoted. Where particular geographical or other details threatened to identify individuals, those details have been altered.
13 / All those interviewed were males – the first women advisory officers were appointed during the period of our research.

these groups expressed an awareness of the need to operate within the institutional mores of the organisation of which they were a part. However, amongst this group were some who espoused (but may not have carried out) alternative forms of practice to that which dominated within their organisation. Alternative structures and organisations were also suggested. Most field staff experienced a tension between their day-to-day work and the expectations, perceived and real, of their organisation and superiors. The third group expressed themselves as the keepers of the institutional memory, as the enforcers and creators of the dominant mores and as those responsible for 'servicing' their political masters. Having said that, it is equally important to realise that each person within any organisation plays some role in this 'culture maintaining system'. This enables those admitted to the 'conversation of the club' to say: He is a good bloke![14] without need for further elaboration. It also enables a senior manager to boast that: 'you know the credibility of our people is enormous . . . it really is enormous and they have the ability to get the message across or to say this is how you should do it'!

A range of dominant themes each with several sub-themes emerges from the interview data. Several sub-themes build up a picture of the first theme: what is involved in the generation of 'valid' extension practice in the rangelands? These include: (i) induction by 'deep-ending'; (ii) the importance of role models for extension officers; (iii) the role of assessment and recognition in determining extension practice; (iv) conditions for the membership of the 'clubs'; (v) the tension that arises between officers' experience of doing extension and the metaphors (and underlying thinking) which is used to describe and manage it and (vi) the gendered nature of extension practice and management. The second major theme is concerned with the pervasive competition amongst the different organisational 'actors'. The third theme – organisation and structures for effective R&D – concludes by addressing a number of sub-themes: (i) the particular structural needs for rangeland R&D; (ii) the question of whether structures, organisation or both warrant change; (iii) suggestions for change and (iv) some political challenges.

5.4 The generation of 'valid' extension practice in the rangelands

5.4.1 *Induction by 'deep-ending'*
Many officers actively sought positions in the rangelands based on prior experience, often as children or through family connections. Some, however, came to a world they had not experienced before. Regardless of any

14 / Bloke – colloquialism for person, usually male.

prior experience few knew what it was to be an advisory officer in this part of the world. Most received no training prior to, or following appointment. One officer described it as being 'just dropped out here and left to my own devices really'. Another: 'I had to set up the agronomy district as I was the first – I lobbed out there – what do I do?' This is induction by deep-ending, a term derived from the experience of being thrown into the deep end of a swimming pool.

The main service organisation, NSW Agriculture, directly employed 200 people (1991–2) in the rangelands, and indirectly supported many others. As sheep are one of the main industries, the Department maintained a number of Livestock Officers who specialised in aspects of sheep production as well as agronomists. Each usually came from a different educational background and were appointed to different Divisions within the Department, which were highly discipline focused (i.e. vets and animal scientists in the Division of Animal Production compared with agronomists, plant breeders and soil scientists in the Division of Plant Production). Few appointees at that time had any explicit training in rangeland science (as described in Chapter 1) or had more than minimal training in extension. As one officer commented: 'the advisory thing just sort of develops as you go along . . . it's very much an individual thing'.

This was also true for staff in the soil conservation service (SCS). As one officer describes it: 'SCS were keen to get me on the right path; a couple of people in particular (graziers) put a lot of effort into straightening me out – [it] tended to get a bit boring with some blokes – they would put you to sleep after a while'. It was not that induction was totally neglected but the dominant advice from superiors and colleagues was of the form: 'get out and meet the rural community – don't try to justify what you are doing just get out there and meet them'. A result for this person was that: 'The first six months I was nearly drowned in cups of tea'. As noted by one middle manager: 'there is a failure in the department – its administrators – and elsewhere to recognise that the true north-west segment of NSW has a climate which is very different to all the rest of it . . .' 'They are also vastly harder than rangelands in America, South Africa and to a certain extent Queensland and Western Australia as well'. The overall effect was to fail to contextualise both the officer and the rangelands.

5.4.2 *The importance of role models for extension officers*

Many officers spoke of the importance of guidance and advice from more experienced officers who acted as role models. Young or newly recruited extension officers actively sought out their more experienced colleagues to guide and advise them on matters about extension and extension practice;

what extension was, what priorities an extension officer should have and how to go about doing extension. Even when the going was tough and an individual found it difficult to stay with the good advice, they used the experience of others as a role model and were explicit in identifying both the role models and the role which they play. In this way particular manners of organisational 'living' were conserved.

Many extension officers talked about the tension that existed between what they found in the field and their experience of the organisation and its procedures. If an individual found that the traditional way of extension was inappropriate to the situation they were in, they also found it extremely difficult to find a pathway or role model which espoused, valued and showed them how to do extension in a non-orthodox way. Those who did try innovative approaches to extension were not seen 'by the troops as role models' or were seen as having 'little relevance to the troops in the future'. During the second phase of our research we experienced this at first hand. One officer, excited at the prospect of working in a different way, had sought contacts and advice ouside of his own organisation. On being counselled to discuss his ideas with one of his superiors, rather than receiving advice and encouragement when he did so, he was 'hauled over the coals'[15] for (i) going outside the organisation and (ii) doing it without prior consultation with his superiors. This was for him a clear message that the Department had its own ways which, it would appear, did not favour unsanctioned outside collaboration or initiative.

The difficulty of course lies in the type of role models available. Officers looked to those who had the respect of the organisation for role models. In so doing they sought as role models those who conformed to what they already saw as a less than appropriate system of extension. This is highlighted in the following excerpt: 'particularly the younger people – they're less experienced ... their role models are the senior officers most of whom work in traditional ways. There isn't (sic) many role models around to demonstrate other ways of working – and I still think we're at a stage where the people we've got are very raw in terms of program development.'

It was a difficult decision for any officer to step out of the traditional role because in all probability they would not get the support of the department. As one middle manager in NSW Agriculture observed: 'The biggest legacy that the name change had – the advisory tag – is that any attempt to step outside the traditional advisory role almost inevitably was met with flack by farmers, other officers and the executive'.

---

15 / Colloquial for censured.

5.4.3 The role of assessment and recognition in determining extension practice

Service institutions all had rigid and hierarchical structures that left many field staff with the feeling of 'looking over one's shoulder'. Officers were constantly and acutely aware of the expectations which the department had of them. They were also acutely conscious of their belief that these expectations were inappropriate in a rangelands setting. As one extension officer observed:

I was also concerned about head office expectations when I was on the downer[16] – I thought they'd look at me out here and say shit he's not doing anything out here – they wanted to know how many farm visits I did, how many press releases and TV interviews – you just didn't do TV interviews out there and farm visits didn't mean anything – it takes you a day to do a visit. I suddenly thought if I am going to be assessed in head office they are going to think . . . he's doing nothing out there and he's not changing practices either.

Such a feeling creates an immediate set of tensions. On the one hand is the desire to be effective within the structure of the department – to attain recognition and to be seen as an able (and conformist) extension officer. On the other hand, is a feeling that such measures of performance are inappropriate to daily experience. Thus the Department is, to some extent, creating the necessity for a choice – success in the Department or effectiveness as a rangelands extension officer. Those two things should be synonymous, but clearly they were not. The dominant theme transmitted to officers was that of the need for accountability to a system which 'counts':

If there was a better way to assess you – that would be good – like how important is the Western Division Newsletter – do we really know? – how important is the local article in the paper? I find the hard numbers approach – you know, how many farm visits, how many newspaper articles, too cut and dried – it doesn't tell you how effective they are – especially out there – you do a radio session over 2WEB and people bring it up with you months later – how do you measure that? – I don't know.

This example raises issues about the methods of assessment. They may be inappropriate for the rangelands, they are also potentially inappropriate for 'extension' anywhere. When the officer says 'I find the hard numbers approach – you know, how many farm visits, how many newspaper articles, too cut and dried – it doesn't tell you how effective they are – especially out

16 / Downer – feeling unmotivated.

there . . . ' he is drawing attention to concern that it is not how much you do but what you do and how you do it that are ultimately important in achieving useful outcomes from extension services.

The real anguish is from those officers who cannot bring themselves to make the invidious choice between the Department's needs and the needs of extension in the rangelands. They wish to succeed in the Department, but they also wish to provide appropriate services to the people in the Western Division. The anguish that this creates is clear from the following extract. The officer was asked: 'If the Department had an identifiable rangeland management service what could it do to better assist those that are working in it?' He replied: 'That's a hard one – recognition is a hard thing – there are not many landmarks you can hang your hat on. It's an attitudinal thing at the higher levels of the Department because they wouldn't see any traditional output from the unit. In terms of what they could do for the people – those out there need continual reassurance about their worth to the organisation and the role they are playing.'

This tension was recognised by middle managers some of whom thought it to be one of the main issues and described it with intense feeling:

I think that the real problem is that our assessment and promotion system has strongly encouraged us to become singularly focused . . . to get on we have to show we have delivered the goods in a specific technical area. . . . If you are going to have a good strong assessment system those same people that are assessing you need to have as a corporate principle – a code of practice which is saying: 'How can we help you?' There is no 'How can I help you?' part of it at all . . . its 'Ah, now we've got an assessment system we can grind these bastards in the ground, sort em out and fix it up.'

Amongst senior managers opinions differed. They ranged from an outright '*No!*' [in response to a question about the need for different assessment systems]: 'There are different parameters to consider and things happen a lot slower out there . . . [but] we're providing a service to them . . . it's the same kind of service' to the acknowledgment that: 'You've probably got to look at different ways of promoting people too – which we haven't done – its one of those things we've obviously got to look at.'

5.4.4 Conditions for membership of the 'clubs'

From the interviews it was possible to discern at least five systems of relationships which have to be managed by field staff working in the rangelands. These were (i) relationships with graziers; (ii) relationships with other rangeland staff, current and former; (iii) relationships with the formal structures of the organization – often one's disciplinary group; (iv) relationships with the 'space' and the bio-physical characteristics attributed to 'rangelands' and (v) relationships with family. Each of these systems of relationships can be likened to a club in which rules and practices exist as conditions of entry and for continuing membership.

5.4.4.1 Relationships with graziers

It was widely accepted that the kind of attributes needed to be an extension person in the west was to be: 'the kind of person that can build personal relationships with graziers and their communities' . . . 'The other thing is you've got to have people who can learn to talk the local language. If you can't learn to talk the language . . . very quickly . . . you're always on the outside.' Advisory officers had learnt from experience that those who succeeded (at least in terms of building meaningful relationships with their grazier clients) were the ones who were able to listen – particularly in the early phases of being stationed in a particular district. One officer describes it thus: 'I used to do a lot of listening – just go around with a grazier for a day and sit in with him while he was going around his water runs'. Another reflected that: 'Bombastic – know all – never let them finish their conversation – just barge in – this is the solution to your problem – often not even finding out what the problem was. I soon found out that wasn't . . . what they [the graziers] wanted'.

The irony is that the experience of not being listened too, of not feeling valued, of feeling that the rangelands were marginalised in the affairs of NSW Agriculture, was pervasive amongst those field staff and middle managers we interviewed. This was not the case in the Western Lands Commission whose *raison d'etre* was to 'manage' [as landlords] the Western Division and the provisions of the state-based Environmental Planning and Assessment Act (but for many the WLC was seen as part of the problem).

Extension officers were often confronted by graziers with the question: 'What bloody use are you?' This question came in many guises and was experienced in different ways in different contexts. Of concern was the possibility that the question could not be answered in ways that maintained all sets of relationships. Many made the choice of 'being accepted as part of the community' as being 'probably more important.' In doing so they ran the risk of playing into a complex set of factors which involved the personal and the political. In some cases this choice could be seen as genuine but perhaps naive. Some officers, perhaps more cynical, perceived that:

individual extension people in previous days have operated in the old paradigm, that empty vessel story, and they are told it is there to be filled up with information. And on the landholders side they have been willing to play that role, probably for too long a period and to accept that role . . . or playing, acting out that role of being the empty vessel and with being supportive to the enthusiastic extension person and then [when] the person leaves probably forgetting 99% of what was said on the day and perhaps saying the guy's a fool or whatever the case may be.

What is not portrayed here are the perspectives of graziers, the other major actor in these relationships (however these are touched upon in Chapters 3, 6 and 8). Officers were aware that graziers made choices to exclude particular people and organisations when they did not share common perspectives. It was reported that one grazier group 'very firmly stated that they would not encompass government departments in terms of the directions of their group – that they wanted to maintain independence from day one and not become an arm of an institution . . . ' Officers were also aware that a background to their relationships with graziers was always the uncertainty in the minds of graziers as to whether those who were responsible for regulation (e.g. the Western Lands Commission, animal health, stock permits, etc.) could also act in their own (i.e the graziers') interests.

5.4.4.2 Relationships with other rangeland staff

The bitter pill for many officers came with the realisation that the only option they had at their disposal in 'constructing' themselves as extension or advisory officers was to regard their work as an art and that like other artisans they had to learn their trade by apprenticeship, to seek out role models. However, as we have described above, the 'successful' role models were likely to represent traditional methods of extension and traditional values. This conundrum was exacerbated by the expansion of Landcare[17] within the region and the recruitment of new staff into the government services who had been exposed to different theoretical frameworks about the nature and praxis of extension and rural development.

The choice, if there was one, was which relationships to maintain, of which of the 'clubs' to belong to: 'there is a fair amount of isolation from the culture of the Department – there is still a great bond between those that are working out there and those that have worked out there – because you form a club – a group of people you can relate to.' To belong meant signing up to a club within a club (existing and former rangeland staff within the overall

17 / Landcare is a network of community groups working together to achieve 'land care' projects devised by them; whilst initiated originally by concened individuals it has become institutionalised through federal government support; at the time of this research Landcare groups were not firmly established in the rangelands where it was sometimes called 'Rangecare' – see Campbell (1998) for background and sources.

Department). A widely held view was that: 'The department never wants to approach the human side of how we work – they never sought out how we saw things, how we viewed the situation – that still prevails in the department – right up through the department. It's never understood that there is a human side to extension.'

Senior managers expressed a particular perspective on these tensions. One, when asked about traditional extension practices and the desire by the executive for the organisation to become more market focused, replied:

I mean they [advisory officers] are good operators and are dearly loved by their farmers, but to try to change them is just impossible. I mean I think we are just going to have to go around them. Even now the younger officers who looked up to these people are now almost despising them. . . . Those old blokes want to watch themselves because if the crunch comes – someone draws the line and says take out everyone below . . . .

Two aspects of this remark are telling. The expression of frustration that officers cannot be 'changed', just as some officers find it impossible to 'change the farmers' and the lack of an apparent ethical commitment to the staff who for so long have been doing what they believe to have been valued by the Department.

5.4.4.3 Relationships within the formal structures of the organisation

As outlined above, NSW Agriculture is constructed around Divisions which emphasise the disciplinary nature of activities – animals, plants, etc. This divisional structure is seen by those in the rangelands as counterproductive as reflected by the comments of one officer with an animal production background:

. . .And also the structure of the Department at the moment with the artificial segregation – of technologies and divisions – one of the first things that struck me out there was the need for me to get a better handle on what the plant situation was – I virtually forgot a vast amount of animal production stuff and got involved with what the pasture situation was because out there it is such a driving influence. [We] inevitably came to the conclusion that the success of work in this area was because we recognised the plant, animal, grazier interaction – this may have come later but it was important to the success of the program – we even drew up a flow chart which started with livestock and ended up with woody weeds . . .

This officer is acknowledging that it is important to meet the graziers on their own ground, that to effectively deal with issues of importance to them, these issues must be set in terms of the daily realities and priorities of the grazier. The structure of the department does not allow this to happen easily. Officers fit into a disciplinary area and are expected to conform to that area of expertise. This makes the provision of a 'whole farm service' difficult for individual officers.

This situation is compounded by the range of organisations which are involved in dealing with farmers in the Western Division (see Figure 5.1) and which result in strained and often competitive relationships between members of different organisations (see below).

5.4.4.4 Relationships with the 'rangelands'

In distinguishing this category we are returning to points made in Chapter 1 about rangelands arising in relationships with various 'stakeholders' rather than existing in and of themselves. What emerged through the interviews were contrasting views of what constituted the rangelands, how they could be described and thus how they might be managed and, flowing from this, what might be the appropriate institutions and forms of organisation. From each perspective a different future form for the rangelands was envisaged; different qualities of relationship also arise.

The main understanding which was contested was whether the semi-arid rangelands was a fragile ecosystem subject to degradation by largely ill-conceived human action (see Chapter 2; Ison, 1993) or whether it was a robust ecosystem capable of adaptive response to human activity. One officer saw 'a landscape which is dynamic and very much alive and in some areas . . . a landscape which is very dead. I've never held with the concept that the rangelands are fragile. I think they are anything but fragile.' From his perspective 'The SCS has created an image of the rangelands as a sensitive environment that is blowing away and they need all this money, National Soil Conservation Project money to fix it up . . . I think that has been magnificently engineered.' Clearly the nature of the rangelands is contested. Such a 'contest' is not unique to Australian rangelands as outlined in Chapter 1; nor is the transformation of the rangeland by European settlement a unique phenomenon in the history of this landscape (Flannery, 1994).

5.4.4.5 Relationships with family

The interviews did not illuminate this set of relationships in any profound way – one is struck more by the absence of the personal – the demarcation of work and home, than of any new insights. What is clear is that for many officers personal considerations were often a driving force for leaving their posting in the rangelands. Alternatively, where partners appreciated the lifestyle it sometimes became the reason for wanting to stay.

5.4.5 The tension between the experience of doing extension and the metaphors used to describe and manage it

Until this moment we have avoided the question of just why traditional extension methods are inappropriate for the rangelands. Let us first say that it is not unreasonable to suspect from the literature that all is not well with extension elsewhere in Australia and beyond (see Chapters 1 and 2). Thus it is not just simply a case of the current models being inappropriate in the rangelands, but rather a case of them being perhaps less appropriate there than elsewhere. What was pervasive throughout the interviews from the 'coalface' to the senior executive was the inability of staff to articulate a theoretical and ethical perspective from which they chose to do extension, assess staff, manage change, etc.

Also pervasive were metaphors which we would associate with the transfer of technology (ToT) and diffusion of innovations models or paradigm (see Chapters 1, 3 and 6). Amongst field staff common phrases and metaphors included: 'improving the knowledge base of you people'; 'learn[ing] it all first and then we can talk [to graziers]'; 'the technologies I have been pushing'; 'it [R&E] means the natural progression of discovering something and getting the message across'; 'information is all out there somewhere. We just have to collect it and put it in a form that is accessible.'

Similarly amongst senior mangers the ToT tradition was alive and well: 'that's really where a lot of . . . technology transfer problems come from . . . we haven't looked and said where does it fit into the business'. It is suggested that: 'their credibility is as extension officers and it's important that they have credibility in advising the business . . . research people are seen as the technology generators and they've got to be seen as credible in technology generation by their clients who are the advisory officers – so if you merge too close, the credibility of both is affected'. For the western division: 'You've really got to design a system to get through but it's got to get through to them [the graziers]'. It is easy to recognise in these statements the linear model referred to in Chapters 1 and 3.

At the time of our research senior managers in NSW Agriculture were enthusiastic that the organisation should refocus around business. Metaphors associated with business were common with them but hardly present in the conversations with middle managers and field staff.

Many of those interviewed knew the language of the alternative paradigm, expressed often as 'co-learning approaches' or 'adult education'. The tensions around the old and possibly new paradigms were summed-up by one middle manager: 'working on the strategic plan . . . recently brought home how strongly entrenched the transfer of technology paradigm is in the Department of Agriculture, and the associated need to set performance indicators in terms of adoption. My suggestion that we should measure our performance in terms of community education was rather summarily dismissed'. This of course raises questions about the commitment to, and understanding of, 'adult education' espoused by senior managers. The following quote from a senior manager suggests that the words may have changed but the underlying paradigm has not: 'The game we are in really is education not advisory-education – so we've got to educate farmers to be better farmers. In the past we have educated not to be better farmers but to grow things better'.

McClintock (1996) researching with service agencies in the UK similarly found the ToT paradigm alive and well, even when the mandate of organisations had extended to concerns with the environment and conservation. Working with the UK Farming and Wildlife Advisory Group (FWAG) he found that the cluster of metaphors around the word 'advisory' revealed a 'message to be delivered'. Under this metaphor Farm Conservation Advisers could be seen as guides, specialists and consultants and farmers as the main clients of the advisory service. What it concealed was the underlying theories of communication (see Chapters 1 and 2), the over 1000 volunteers associated with FWAG and considerations of who pays for the advising. His research focused on the role that metaphors play in revealing and concealing particular understandings from which our actions arise.

When given the opportunity to reflect on their practices (via the interview process) some officers recognised behaviours which were in marked contrast to the 'changing them' and 'transferring information' paradigm. One officer reflecting on his conversations with graziers observed:

They never ever confronted you directly with questions – they worked their way around you – get your measure and then they start to sneak in some things ... they didn't want to invade you ... but they wanted to know something ... you could tell – so you just listened and let it happen. In a lot of cases you ended up feeling as though you'd only confirmed in their mind what they'd been thinking – they seemed happy with that and I think it helped a lot'.

Some of the implications of this insight for the design of R&D systems is taken up in the final section of this chapter and in Sections 3 and 4 of the book.

5.4.6 *The gendered nature of extension practice and management*

All staff interviewed in the initial phase of our research were men – at that time there were no women advisory officers or managers. Not surprisingly the topic of gender imbalance amongst advisory officers was rarely introduced into the conversation. When it was introduced by a senior manger it was done so in the following way:

that's why we're going for this Women's Rural network because that's the first opportunity we've ever had to target women . . . we haven't been able to target women – really – we really can't go out and target housewives .. .we've hoped they would come along to these other things. Farm business management – Farm Cheque[18] *. . . that's where the action is for rural women'.*

This statement reinforces research by Margaret Alston (1998), who relates very powerful stories of 200 years of excluding women's voices from all aspects of Australian agriculture except the hard work; whilst the institutional context is changing and the forums in which women's voices are being heard is increasing, her research, and the perspective provided by this senior manager suggests that this is not happening quickly enough.

In the field, advisory officers were often disconcerted by the need to relate to women; they seemed ill-prepared and uncomfortable. As one observed:

your work was done around a cup of tea at the kitchen table . . . and although you would be directing your conversation to the husband, the wife would be on the edge . . . you were very conscious they were there and often they would join in. It depended on the nature of the woman herself – some were very aggressive . . . some worked on the farm as hard as the husband.

Another officer contrasted his experience in the rangelands with his earlier experiences in the main agricultural areas:

When you went there too, I hadn't struck it before so much in the temperate regions . . . when you sat down for a cup of tea and did your business the wife was always there, always . . . I mean she could have been doing the washing but she always came in and participated – I think that indicates how important the wife is in the economic management of the place, even with fencing, mustering or whatever.

As Ison (1993) observed, gender and ageism, or its reverse, become critical in this context – decision making and thus agreement about what is desirable change has for too long in rural Australia suffered from the lack of full and open participation by women. There is also the risk of constantly making decisions which are generationally skewed as those in positions of power draw on experiences which are past and no longer relevant to the

18 / A farm management programme run by NSW Agriculture.

experiences of the present. It seems unfortunate that young people who might one day manage the property were often excluded from the conversations between graziers and advisory officers. As one officer observed: 'on the other hand if they had sons [sic] they generally didn't sit in – I found that . . . the boss, the father would let the sons go out fencing or whatever and he and the wife would sit down and talk'.

5.5 Rivalries which interfere with cooperation

Competition between personnel or organizations was mentioned in most interviews with field staff and middle managers. This took a number of forms: (i) rivalries between staff in regions or divisions within an organisation; (ii) rivalries between researchers and extension officers; (iii) rivalries between members of different organisations providing services in the Western Division and (iv) rivalries between organisations. This is not new and has been commented on by past inquiries into the 'future' of the rangelands (e.g. the Fisher Inquiry). However, it persists.

Many extension officers commented that there is 'little co-ordination between agencies – we get on well but tend to work on our own' and the 'strong competition between Western Lands and SCS and Ag[riculture] . . . that was pretty well across the west.' It was acknowledged that initiatives such as Landcare were changing this, although organisationally the SCS tended to take a proprietorial interest in it. This led officers to suggest initiatives in areas where other organisations had less influence, or in the move towards group-based extension, to become possessive of 'their groups'. For many NSW Agriculture staff the 'SCS [was] always pushing tractors and bulldozers (a reference to the hire scheme for machinery which historically has been an important feature of the SCS operation in NSW). For officers outside NSW Agriculture it was sometimes difficult to see a continuing rationale for its presence when its main concern was production – given that on a state-wide basis production in the Western Division was relatively low. Attempts at collaboration in advisory functions sometimes foundered on unequal contributions and interpersonal rivalries. The story for research was different, with evidence of collaboration between members of different research organisations.

It is telling that collaboration was healthier for research than extension; this is undoubtedly related to the perceived and often real competition that existed between researchers and extension officers. Sometimes this was personally motivated (as, for example, by failing to invite extension staff to participate or join in, either deliberately or inadvertently) but was generally institutionalised through the different reward and recognition systems. There was no history of doing 'extension research' and the extension officers 'don't see themselves as researchers – they see themselves as advisory officers who get told answers to problems.' Some felt it was changing; one manager commented: 'when I first went into regional management – what I saw was straight agro [aggressive behaviour] and competition between research and advisory leaders – they were all out for their dollar and stuff the other'. This person now considered that there were more teams forming 'at the front line', although intriguingly, given the substance of our interviews, one senior manager felt that 'the people out there [in the rangelands] now are the closest thing we've got to almost a programme approach now – they tend to work in a team'.

Middle managers, particularly those with a regional perspective, found competition frustrating at a number of levels; if you were concerned with rangelands then there was 'competition with cropping areas' staff'. Or there were 'divisional barriers in this; at the coal face divisions often do not exist but structurally they interfere'. In some instances rivalries were exacerbated at the level of regional management, and it was never clear how collaboration with other agencies in advisory services was viewed from head office. This lack of clarity enabled observers to claim that particular organisations were in competition in the rangelands.

5.6 Organisation and structures for effective R&D

5.6.1 *The particular structural needs for rangeland R&D*

Whilst the service organisations seem to have no vision for extension in the rangelands, they nevertheless see a necessity to maintain a presence there. Perhaps this is a politically motivated need? None-the-less it is a presence which is managed and judged, in the case of NSW Agriculture, in much the same way as all other areas of the department's extension activities. There are a range of structural issues which contribute to a view of a poorly fitting extension model. Perhaps the most dramatic of these are to do with the natural cycles of the rangelands, their productivity and the persistent nature of the animal and pasture production problems.

This scenario is exemplified by one set of reflections:

I lost enthusiasm for a while – I thought . . . it's going to take a long time to change people's practices – I went through a bit of a downer for a while – I would say to . . . my wife, this is going to take 4 or 5 years to achieve anything here . . . I do think you need research and extension services out [here] *because we haven't got all the answers and in* [this] *country the decisions . . . made about grazing management can affect you for the next 10 or 15 years or longer – unless they* [the graziers] *are given the right information and use it in the right way they can affect their long term viability.*

Structural impediments to their on-going effectiveness such as this often led officers to feel they were in the wrong conversation.

5.6.1.1 In the wrong conversation?

The constraining structures are made further apparent by one officer who found in-service training irrelevant to his needs:

It was always very frustrating to me – I'd go away to the normal in-service conferences – they were just foreign – by the third day of the conference you would be sitting there thinking what the hell am I doing here – it was like the bloody dark side of the moon – and that only served to highlight the gap – while you were included in the conferences to make you feel part of what was going on, they only served to highlight the gap.

This comment raises many issues. It highlights the tension which inappropriate departmental expectations create, whilst at the same time raising the spectre of a department which, at the top, is seen to be out of touch with the realities of the rangelands. Indeed the speaker goes on to indicate that the senior ranks of the department have no vision for extension in the rangelands. More interesting from the personal position of extension officers is the expression of a feeling of separateness – the officers working the rangelands felt that their territory and its needs were so different as to bear little relationship to other extension territories. There was a widely held view amongst field staff and some middle managers that new structures or a new organizational form was needed to operate in the rangelands. We return to this below.

5.6.1.2 Conflict between short-term political demands and long-term production and bio-physical processes

We have already made reference to the mismatch between assessment and promotion processes and the 'realities' of working in the rangelands where change is a long-term process. This tension is also manifest in the mismatch between short-term political demands (the State of NSW has elections every three years) and the needs of programmes which are by their bio-physical, environmental or commercial nature very long-term.

For example, one middle manager related his experience of one 'high profile' programme:

the biggest difficulty I had with that job – was the politics – there were unreal political expectations – and the top down influence was very strong

... [there was] *no real appreciation of what was happening on the ground'. 'What really led to the difficulties is that the problem of a government with a short-term political agenda being able to sit comfortably with the long-term nature of the programme; I mean the long haul involves huge change over 20 years – not 2–3 years.*

Despite these difficulties he still considered that 'obviously we have to accept political priorities'.

Field staff expressed extreme frustration when they perceived graziers doing the right thing in their management practices only to find this counting against them when dealing with financial institutions who also took a short term view:

You know [one farmer] *– the RAA* [a rural adjustment lending scheme] *knocked him back on the basis . . . that he must be overstocked, otherwise he wouldn't be pulling scrub*[19] *– now I know . . .* [that property] *is running less sheep than ever before and the only reason he is pulling scrub is because he has a long term plant community management strategy.*

Because of the slow pace of change even subtle changes in management have far-reaching effects on the long-term viability of graziers. The nature of the ecosystems and the complexity of the interacting factors mean that a change made now may severely damage the viability of the holding far into the future, and when the damage becomes obvious it may be simply too late to take corrective action. As one interviewee observed:

The one example that really drove that home to me was, one fellow who has tended to try to do the right thing – he has a property in that rosewood/belah[20] *country and he had a paddock he considered he had given a bit of a caning* [subjected to heavy grazing] *over the previous 3–4 years and decided he needed to give it a spell* [a rest from grazing] *and so he gave it a spell – it turned into solid woody weed. Obviously it was a pretty bad decision but he was taking on board a general theme of advice in terms of what was considered to be the right thing to do for that sort of country and it turned out extremely bad for him. He made his decision at the wrong time . . .*

This highlights the problem of simply having to wait till the right time to take any management action at all a theme taken up in Chapter 1 in the discussion about models of vegetation dynamics. Indeed it may be that the

---

19 / Scrub – dense, low growing native trees comprising usually a mix of species and edible by sheep when cut.
20 / Rosewood and belah are two native tree species which grow in association. The term is used to describe a particular type of landscape.

implementation of an initiative such as burning woody weeds must wait five years or more before there are the right conditions for its implementation.

For others there was a mismatch between the organizational and structural arrangements for moving towards a sustainable agriculture: 'we think we are on top of sustainable agriculture because we put out a folder with 10 ... strategic plans in it – it's more complex than that' ... 'You've got to start from the premise of what is agriculture going to look like in 10 or 20 years time – what are the issues -what are going to be the gaps – where is the role of government, which [activities] are public good – what are private good?'

5.6.2 *Changing structures or organisation?*

What would we ideally design as a new way for NSW Agriculture, or other service organisations, to operate in the future? Such a question is by definition the wrong question. It presupposes that the needs of existing organisations are those which we wish to meet. Indeed it presupposes that we agree on the continued existence of these organisations and *de facto* to their continued existence with structures similar to that which exist now. None of these assumptions should be accepted. We choose to go beyond the very pertinent question asked by one senior manager in NSW Agriculture: 'you've got to ask the question – what is our core business? – I have a lot of trouble answering that question – it's not clear cut by any means.'

If we accept that farmers' needs are tightly linked to their social context, are knowable to the farmers themselves and are only peripherally technical, then we can argue that any future relationship between a government organisation designed to work with farmers, and farmers themselves, must possess certain fundamentals: the most important parts of the relationship include an acceptance that knowledge and the possibility of triggering new understandings is the main basis for such a relationship; that need is transient, contextually bound and often complex; that farmers are the only people who can come close to deciding how such a relationship should best be implemented and what it should do, and that farmers are the people who must take control of any such relationship; that the current sets of relationships are based on a view that farmers have needs which are often unknown to them and that departmental officers have the answers to these needs and that they must provide these answers, resulting in a fundamentally hierarchical relationship which disempowers the farmer; that such a relationship is not supportable or appropriate and that such a relationship will inevitably be only a part of the fabric of a wider and more durable network of relationships.

Graziers, like the rest of humanity, are fundamentally social. They require to be part of social relationships, and these relationships often involve the sharing of stories, anecdotes, knowledge and feelings. 'Agricultural information' – for lack of a better title – is nothing more nor less than a subset of this social fabric. Notions of 'formal' and 'informal' information are not only irrelevant but perhaps entirely unhelpful. Research work into rangeland productivity appears to point to subtle managerial decisions having long-term productivity impacts. These subtle decisions are far more likely to be exemplified in anecdote and story than in research papers and information brochures. Thus the network of social contact is perhaps the most important path for the emergence of new knowledges and practices in such an environment. It is against this set of perspectives that we choose to examine any strategies for change.

Our research has focused on the advisory or extension staff, but as has been argued elsewhere (Chapters 1, 3 and 9) to distinguish between research and extension has questionable utility in the context of agriculture and rural development. The research by Kersten (1995) (who examined the experience of researchers operating in the Australian semi-arid rangelands) and our own clearly shows this view is contested and certainly would not have been acceptable to the senior managers of the service organisations at the time they were interviewed. However, all those we interviewed recognised and accepted the need for change in the provision of R&D services in the semi-arid rangelands. What they did not agree on was what to change or even the nature of the change process itself. Many of the suggestions for improvement that were made by interviewees were concerned with trying to make the existing arrangements work better – structural or first-order change. There were those however who recognised the need for fundamental change – change which changed the whole system and their place in it, both individually and organisationally (i.e new organization or second-order change).

To achieve meaningful first order change in order for any extension service – as they were then constructed – to function effectively there needed to be widely held and accepted answers to two sets of questions. Firstly 'Why are we doing extension out here?' Extension must be seen as relating to a particular view of a situation, to the maintenance of an agreed set of relationships and contributing towards an accepted set of goals. These requirements are evident on two levels. At the first level is the legislative and policy goals of government. The western division is 'managed' by several government organisations, each of which should have goals and requirements for the management of the area and programmes designed to meet those goals, whether they are about the biological management of the area or the 'welfare' of individual graziers. The second level is that associated with what could broadly be described as the goals of the graziers both individually and collectively. Interestingly extension officers often referred to their belief that this second area was where they needed to derive their direction from if they were to be effective. It is inevitable that government employees must contribute to the achievement of governmental and departmental goals, although this begs the question about the process of their formulation. Any other course would ultimately prove unworkable. However, in the absence of any governmental vision, officers are left to decide on their own extension direction and to evaluate a complex set of factors in doing that. The outcome is that officers feel that they must get on as best they can, whilst feeling that there is an effective policy vacuum.

The second set of questions can be summarised by saying 'If that's why we're doing extension then what should we be doing and how is it best done?' Often the only part of these two questions that ever gets asked is 'How can I do extension better?' In the absence of a sensible framework of goals and directions, there is no useful answer to that question.

Within a first-order perspective a 'solution' which is suggested from our interviews is to: (i) push for a political statement of vision for the rangelands against which (ii) a strategic and operational plan can be developed, which (iii) encompasses measures of performance against which activities and staff can be assessed and which (iv) is conducted in new structures (e.g. a unit or centre or institute, or assessment procedures), which better meets the needs of operating in the rangelands.

Such a process might be achievable and result in an improvement. We, however, are sceptical as are some of those we interviewed. As a process it would be open to formulating a system which might be described as either: (i) a system to enhance the power of organization X or (ii) a system to woo the rural voters of NSW so that our re-election can be assured. Our interviews suggest that from certain perspectives both 'systems' exist or have existed. One interviewee involved in a high profile, politically motivated programme characterised it as: 'the other thing that really stood out is that a lot of the staff we were required to organise – had that high flying high profile approach . . . which is fine – successful for getting it on the agenda but little or no grass-roots consultation'. However, he considered the absence of grass-roots involvement a threat to success in the long-run.

There is clearly a crisis of purpose amongst all 'stakeholders' in the semi-arid rangelands of Australia and there are no structures capable of orchestrating an on-going conversation about what a future rangelands might be. This is not a recent phenomenon – it has been a recurring theme since European settlement as described in Chapter 4. What is comparatively new is the number, the 'density', of the current set of service organizations which continue to compete and confuse (Figure 5.1) and the extent to which understandings in service organisations 'spill over' into other organizations. For example: 'People will still try to make a living there but the reality is there is no clear strategy in Government about those areas – the home maintenance concept has been done away with but even financial institutions have an inadequate understanding of what is required out there'.

5.6.3 *Suggestions for change*

The resistant discourse (following Foucault, 1985) that emerges from our interviews is of two forms. The first is that present and past organisations and institutions create a dependency culture in a rangeland that is 'over administered . . . in many ways the Western Lands Commission has become a crutch for those people' . . . 'so it's constantly used to distort their view on government administration – it's also an added cost and it's just inefficient'. One experienced manager commented:

It seems to me that the executive of past commissions and its staff have all very much fostered and developed the view within the pastoral industry and the landholders out there that we are essential to you – if we go you're going to get done – and the classic example is that today the state planning and environmental act is used to scare hell out of them – where[as] *you could have regional environmental plans where parameters could be set* [by relevant stakeholders] *for the conduct of agricultural operations and then* [the] *regulatory side would be simply to assure that people did not go outside these parameters.*

This potential 'system' envisages participation by a broader but undefined range of stakeholders but certainly graziers. It involves the maintenance of organizations not unlike the current ones, but including one specifically mandated to service the whole of the rangelands: 'you need . . . a Manager of Western Lands Operations who has responsibility for the totality of the thing – he knows what his allocation is and his staff and so forth and he can get on with training and programmes and so forth'.

What is not clear is whether such an organisation would continue to house regulatory and advisory or 'developmental' activities under the one roof – a problem which leads to conflicts of interest and perception (and which is common to many Ministries of Agriculture, e.g. Britain and the so-called 'beef crisis'). What is also not clear is what structures would be needed to break out of the dependency which was of concern to this commentator.

Interestingly a senior manger in NSW Agriculture perceived the need in terms of:

joining up [staff from research and extension] *problem definition . . . If the problem definition is done and recognised by both you get the right answer.*

Part of the problem with technology transfer is the research blokes identify a problem which the extension blokes don't see as the problem – then they find the answer to that problem and the extension blokes say what bloody use is it . . .

From this perspective it would seem that other stakeholders are not to be admitted to this process nor is there any awareness of what might be required to institutionalise such a process in ways which did not perpetuate dependency. Robert Chambers (1997) describes it as maintaining the 'power of uppers'. Research reported in Part III of this book provides one model for breaking out of dependency.

The second resistant discourse to emerge from our interviews was more radical in nature. It was expressed in a number of ways: 'the general feeling amongst landholders (and some agency) staff is that extension agencies need to change to a more co-learning forum than that in which they currently exist' or 'I think we have got to try and get people together with similar concerns – now I don't think government and bureaucrats should be telling people what they should be concerned about' and 'that means you have got to understand relationships'. In addition to the CARR project (around which this book is based – see particularly Chapters 6, 7 and 8) other independent initiatives were occurring which were attempting to give form to this alternative R&D system.

Triggered by experiences of community collaboration in the control of rabbits, and through the increasing availability of Federal funding for group-based community action (associated with the national Soil Conservation Programme and Landcare) one officer had facilitated the emergence of a number of landholder groups. At the time of our research he and group members were in the process of exploring how to generate synergies via an umbrella organisation. He had been aware that in the US, where land-holder groups receive federal funding, they 'actually employ to some extent government people to do work for them on behalf of their groups'. This of course changes power relationships through changing the nature and capacity to be responsible. He observed that: 'the more I thought about it and the more I got into it, the more I realised that it was probably institutions that lacked . . . understanding and not the landholders'. From this person's perspective what was required was to 'create a planning environment as an institution rather than spending so much time and energy trying to create institutional plans which other people are supposed to adopt' and a system which made possible 'community learning.' A learning cycle which included feedback was seen as central to achieving this as was the need for group members to develop 'negotiating tools' and 'mechanisms by which they [groups] can . . .

either change or die.' Underpinning his enthusiasm for this model was the belief that 'the institutions here are reacting to changes that are occurring through this process and other processes from other areas. I don't think they necessarily lead the process.'

It is possible to speculate that the vision of an alternative model described above was only possible in a context that was relatively free from bureaucratic intervention (due to isolation) and in which opportunities emerged due to the changing policy and funding context at Federal government level. What seemed apparent was that this initiative was free from much of the 'gatekeeping' activity that some organisations exercised over Landcare (see Campbell, 1998) and other group-based initiatives in other areas of the state. Structurally, however, it was always open to 'sabotage' via these tendencies whether overtly or inadvertently.

Interestingly this model incorporates several of the features identified in a comprehensive study of African pastoral organisations. Sylla (1995) arrives at ten conclusions which could be considered, critically, in any process of re-design of organisations and institutions for pastoral regions. Clearly these are not, and should not, be considered as directly transferable to NSW. Yet they provide insights and resonate with initiatives that were already beginning. They are: (i) use ad hoc organisations; (ii) ensure membership is flexible so that long-term solidarity is enhanced over short-term benefits; (iii) support bottom-up and top-down approaches which are complementary; (iv) support small organisations; (v) support weaker groups; (vi) take into account traditional systems; (vii) do not focus on one strategy or group in isolation; (viii) use flexible planning which is iterative and adaptive; (ix) support privatisation and collectivisation with caution and (x) support decentralised authority.

The proposal of a new form of organisation also resonates with the experiences of Margerum and Born (1995) in their case study of resolving disputed landuse in the Lower Wisconsin River valley in the US. They achieved a significant breakthrough when a 'common overarching goal' – to preserve the existing character of the region – was identified amongst a majority of often competing interests. A new organisation (and structures) was, however, needed to achieve this. They also noted that 'the more complex and controversial aspect of the process was reducing this broad consensual goal to specific management objectives and actions' (p. 382).

The process employed by Margerum and Born (1995) and which is apparent in the alternative model proposed above, in some Landcare groups and in some Agenda 21 groups (Open University, 1997) can often be considered as purposeful systems because they aspire to the same overarching goals but do so in many different ways, and where successful they adapt and change their goals as they learn from experience through a self-producing network of conversation (see Chapter 2). This is very different to the situation where an organisation and its senior management are effectively there to ensure the continuation of the organisation in its present, or a comfortably modified, form; where the original reason for the existence of the organisation is paid lip service but in fact its role appears to be to nurture itself rather than, as might be assumed, the network or relations which constitute it as an organisation.

5.6.4 *Some political challenges*

Some of the political challenges faced by alternative and complementary models for R&D in the NSW context are summed up in comments by senior managers of NSW Agriculture: 'it's those sorts of comments plus some more dubious comments [*re* failure of technology adoption] that have come from consultants with some vested interests in mind – that they are trying to put across the story [that] the public sector has failed in its job of technology transfer. Really, at the end of the day they are looking at a job for themselves.' Or, an emphatic 'No!' in response to the question of whether different staff are needed in the rangelands? 'It's still a business – as long as we have that focus' and where their vision is that: 'the type of extension we have to do [is] – getting farmers to think about what happens to their product beyond the farm gate – getting in touch with the market – getting them to run a more commercial market driven business that's no different to Dalgety's, David Jones [major Australian firms] . . . it's about farming as a business' and that the core business is 'the food and fibre business with a focus on the business of production, selling marketing, processing, manufacturing, consuming and exporting'.

These sentiments were later institutionalised as an outcome of a major strategic review of NSW Agriculture, entitled a 'fundamental review', underway during our research. The core business was redefined as: 'The Department is primarily a service provider to the food and fibre industries and functions as a link between consumers and primary producers. This is achieved through the provision of integrated and market driven policy, research, advisory and regulatory services'. What is not clear is whether in redefining 'the core business', it was envisaged that this was going to be achieved by changing structures, or whether this constituted a new 'organization'.

Because the senior management are engaged in a pursuit of resources (research funds, etc) and a protection of the status quo, long-term vision is therefore subsumed within the more immediate demands of politicians and lobby groups. Any suggestion of some sort of philosophical basis for the work of the organisation is met with stock phrases and out-of-date ideas. The critical issues for this group are those associated with fulfilling the political agendas of their various masters. But two stories told during our interviews suggest that this is potentially a trap in the long-term.

The first story reflects the crisis of purpose service organisations face, particularly in terms of the trade-offs to be made in activities which can be seen as either a public or private good:

take OZPIG [a decision support system] – it's a classic – I mean OZPIG now ties all the biology together with all the dollars and so you can change any bit of the biology, the genetics, the feed, the environment and you can look at the bottom dollar. It's probably a bit sophisticated for the majority of our pig producers but its not too sophisticated for the people who produce the majority of the pigs – 30 percent of producers produce 85% of the pigs – the other 70% I don't think have the ability yet to take on that technology – you could also say – why worry about them anyway if they only produce 20% of the pigs. Unfortunately though we have most contact with the 70% who produce the 20% rather than the 30% who produce the 80% and that's a bit of a quandary for us – I mean we really have difficulty getting the foot in the door with the big guys'.

Having redefined its core business as above, NSW Agriculture still had no basis on which to decide if it was going to continue to work with the 70%, who clearly need services other than that provided by OZPIG,[21] or the 30% of 'high fliers' who would receive their service free.

The second story concerns the question of whether in the context senior managers find themselves in, it is possible to 'manage' in any rational, goal directed way to achieve any sense of purpose that they might articulate (by whatever means). The anecdote was given in response to the question as to why NSW Agriculture has escaped any significant change compared to similar organisations in other states and other countries:

'I think it was just the string of 10 years of almost fortuitous events that allowed us to survive rather than anything that we did – just luck. I think we went through a period when we had a Labor party in power who believed they had to have the country vote – therefore they looked after agriculture reasonably well – our budget wasn't cut in line with other public sector type

21 / Australia suffers at state and federal level from not having any broad-based rural development strategy and the organisations and structures to pursue this. Unlike the UK and other parts of the world there is a paucity of strong non-governmental organisations and community-based organisations concerned with the 'countryside' and rural development. Landcare may change this but has not at the time of writing.

departments – we then had a change of government and a consulting report indicated that tourism and agriculture were the two best ways that the state could utilise to trade itself out of trouble – and it was that report that allowed us to escape budget cuts for the next three years of government'.

5.7 Concluding remarks

Extension officers in the semi-arid rangelands of NSW are constituted through a particular network of power[22] which includes theoretical models, particularly the ToT model, institutions conceived as particular practices or the 'rules of the game' (e.g. promotion criteria), and the experiences of the extension officers and managers, particularly in terms of the choices they make in developing and maintaining particular relationships (e.g. with graziers, other officers as role models, etc.). Extension and technology are, in the eyes of service providers in the NSW Western Division, inextricably linked. The practices of R&D professionals is to undertake research or administer the land, thus developing new technology which maintains technological lineages (see Chapter 4), and subsequently to extend that research to growers or to police the regulations. Through these practices, and the practices of other stakeholders, the rangelands as we know them are 'constructed'. In this network of practices there is limited opportunity for new understandings and practices to emerge because of the 'rules of the clubs'. Debate does not occur on the central issue of extension as technology transfer (see also Chapters 2 and 3). What is more problematic is that the nature of change itself is not considered and conversations become impoverished because of this and the failure to distinguish between organisations, institutions and structures.

As always there is resistance. Seeds of change exist, but whether this will be first- or second-order change remains to be seen. It is clear that current organisations and structures are inappropriate for the effective conduct of rangeland R&D. It is hard to conceive that this will change in the organisation that is envisaged by NSW Agriculture's new mission statement – not because the focus on business is necessarily irrelevant but because of the concommitant need for a change in structures to achieve a change in organisation and because of on-going contingent management in response to short-term political imperatives. What is missing is a theoretically and ethically valid framework from which to build relationships with major stakeholders in such a way that the resulting organisation was adaptive to the bio-physical characteristics of the semi-arid rangelands and its associated human activity.

22 / See Chapter 4 where this analytical framework is developed.

References

Alston, M. (1998) Women, the silent partners of agriculture. In *Proceedings of the 9th Australian Agronomy Conference*. Australian Society of Agronomy, Wagga Wagga, Australia.

Beck, R.F. (1991). *An assessment of New South Wales Agriculture Involvement in Current and Future Research and Advisory needs on New South Wales Rangelands*. Mimeo Report to NSW Agriculture. 18 pp.

Blackburn, J. and Holland, J. eds (1998). *Who Changes? Institutionalizing Participation in Development*. Intermediate Technology Publications, London.

Campbell, A. (1998). Fomenting synergy: experiences with facilitating Landcare in Australia. In *Facilitating Sustainable Agriculture. Participatory Learning and Adaptive Management in Times of Environmental Uncertainty*, ed. N.G. Röling and M.A.E. Wagemakers, M.A.E. pp. 232–49. Cambridge University Press, Cambridge.

CARR (Community Approaches to Rangelands Research) (1993) *Institutions Responding to a Community Request: Wool Marketing Advisory Support Group . . . and Beyond*. Monograph. University of Sydney and University of Western Sydney. 48 pp.

Chambers, R. (1997). *Whose Reality Counts? Putting the First Last*. Intermediate Technology Publications, London.

Flannery, T. (1994). *The Future Eaters*. Reed Books, Sydney.

Foucault, M. (1985). *The Use of Pleasure. The History of Sexuality. Vol. 2*, translated by Robert Hurley. Penguin, London.

Ison, R.L. (1993). Changing community attitudes. *The Rangeland Journal*, **15**, 154–66.

Kersten, S. (1995). In search of dialogue: vegetation management in western NSW, Australia. PhD Thesis, Department of Crop Sciences, University of Sydney, Australia.

Margerum, R.D. and Born, S.M. (1995). Integrated environmental management: moving from theory to practice. *Journal of Environmental Planning and Management*, **38**, 371–91.

McClintock, D. (1996). Metaphors that inspire 'researching with people': UK farming, countrysides and diverse stakeholder contexts. PhD Thesis, Systems Discipline, The Open University, UK.

North, D. (1990). *Institutions, Institutional Change and Economic Performance*. Cambridge University Press, Cambridge.

Open University (1997). *Environmental Decision Making: a Systems Approach*. The Open University, Milton Keynes.

Röling, N.G. and Wagemakers, M.A.E. eds (1998). *Facilitating Sustainable Agriculture. Participatory Learning and Adaptive Management in Times of Environmental Uncertainty*. Cambridge University Press, Cambridge.

Swift, J. (1995). Dynamic ecological systems and the administration of pastoral development. In *Living with Uncertainty. New Directions in Pastoral Development in Africa*, ed. I. Scoones, pp. 153–73. Intermediate Technology Publications, London.

Sylla, D. (1995). Pastoral organizations for uncertain environments. In *Living with Uncertainty. New Directions in Pastoral Development in Africa*, I. Scoones, pp. 134–52. Intermediate Technology Publications, London.

Wright, S. (1992). Rural community development: what sort of social change? *Journal of Rural Studies*, **8**, 15–28.

Part III
A Design for Second-order Research and Development

This part is concerned with the development and application of second-order R&D in the semi-arid rangelands of western NSW. Our aim was to generate, through our research, new understandings which might guide the design of future R&D systems. We did this by 'braiding' theory with subsequent practice in an attempt to create a context in which second-order R&D was possible. From the start we did not set out to pursue a 'fact finding' mission but rather, via the medium of 'stories', to invite pastoralists (graziers) to tell of their experience.

Our aspiration for our second-order R&D was for practical outcomes. As described in Chapter 1, our design was based on an increasingly recognised need for emancipation from dependency, empowerment through collaboration and collaboration based on the mutual acceptance of different realities. Based on our earlier experiences we postulated that the joining together to act based on shared enthusiasms all flow from doing concurrently and with awareness first- and second-order science. The research we describe in this section was designed to test and model the tenets of the complementary nature of second-order R&D.

As researchers beginning in 1990–91 we had no experience of putting these concepts into action in such a setting. Neither was there a history or tradition of this type of research on which to draw nor a network of like-minded researchers to act as guides and mentors. When we embarked on the project we had developed the following approach: families of pastoralists were invited to tell of their day-to-day experience and where possible, their understanding/interpretation of their experience. Semi-structured interviews that were used to trigger accounts of their stories were designed to map out patterns of meaning across time: first, the historical context; second, the present-to-hand experience; and third, the anticipated context (the future). Our intention was that this phenomenological data (data based on experience and action) would be coupled with hermeneutic data (how the family members make sense/interpret their experience) to constitute the contextual research focus (second-order data). The mapping of patterns of analysis constructed from social, ecological, and pastoral events (first-order data) proceeded along a parallel and dynamically interrelated path, although as described in Chapters 3, 4 and 5 the data we generated was not typically ecological or biological – our conclusion was that many others better equipped than us were already doing this.

Our research involved attempting to hold first- and second-order approaches in a complementary tension. Based on this awareness the participants were invited, in specially designed workshops, to identify their enthusiasms for taking action in particular domains (social, political, flock management, etc.). It was envisaged that given that these people shared a common geographical area, there would be some groupings formed along the lines of shared enthusiasms. These groups were to constitute 'user-initiated R&D groups' responsible for the generation, management, and subsequent evaluation of actions designed to benefit themselves as a pastoralist community.

In designing our research we attempted to be consistent and thus rigorous in our 'braiding' of theory and practice (Table III.1). The following three chapters outline what we mean by this and is an account of how our practice dovetailed with our conceptual modelling in an early phase of our research work (Table III.1).

Our research was an informed (theory-based) attempt to:

- Accept pastoralists (all involved family members) as competent researchers in their own right.
- Integrate their enthusiasm for 'research' with maps (patterns over time, space, decision making) of economic, social, political, ecological, and range management data.
- Reflect back to the pastoral community, the value of their 'traditional' R&D knowledge for the sustenance of this way of life.
- Articulate the theoretical underpinnings of this research approach for the benefit of the broader scientific community.

In the first chapter of this section (Chapter 6) we describe research based on an experiential model of doing science in which we tested our concept of 'enthusiasm' as an emotional predisposition to action which might act as a basis for shared R&D action when triggered. We know of no similar research in the literature. Chapter 7, written by Lynn Webber, who was one of the research assistants on our project, grounds our research approach by providing some of the detail of how we went about doing our research with grazier families in the Western Division of New South Wales. Webber stresses that this is not a prescription for others to follow but a story of one example of second-order R&D which might trigger insights for other researchers. Her aim is to give some insight into how we developed our research framework using theoretically based principles to guide our action.

Table III.1
A conceptual framework based on the braiding of practice (the experiential world) with theory (the conceptual world) on which our research design was initially based

Experiential world		Conceptual world
Invitation & semi-structured interviews. Stories of experiences (present, past and future)	→	Enthusiasms for action are elicited through showing 'acceptance'
Analysis & mapping of enthusiasms using pastoralists' own words Invitation to attend group discussion;	→	Researchers are both catalysts and conceptualisers
Invitation to be co-researchers; Presentation & discussion of maps; Invitation to generate afresh enthusiasms for action	→	By becoming co-researchers pastoralists assert 'ownership'
Pastoralists nominate the issues **they** want to act on	→	Pastoralists act as co-researchers and share direction of research
	→	Participatory research design is now in place

A further rationale for her chapter is that we wish to be up front about the work, sometimes anguish, and struggle that we have experienced in attempting to carry out our research. The question is posed: What are some of the ways in which we have worked that have been helpful? We do so because we have no desire to sanitise, totally, and provide polished claims.

The final chapter (Chapter 8) in this section describes, from the perspective of some of the graziers involved, the outcomes of the collaborative, second-order R&D that was conducted over a two-year period. The introduction to the chapter was written by Sandy Bright, one of the grazier participants in the research. The remaining material was prepared by Danielle Dignam and Philippa Major, also research assistants on the project, based on the contributions of Sandy and Cleve Bright and Kim and Margot Cullen, graziers who responded to our invitation to prepare a chapter based on their experience.

6 *Enthusiasm:* developing critical action for second-order R&D

David B. Russell and Raymond L. Ison

6.1 Introduction

We were on a dust choked road somewhere between Wirrinya and Forbes in central western New South Wales. It was getting towards the end of an intense but absorbing week in which we were conducting the first ever 'rapid rural appraisal' (RRA) in Australia. This had involved visits to many farms and spending time with the farm families to hear about their histories of farming and the concerns and issues they now confronted. The impetus for the RRA had come from concerns about how and why particular research questions were asked and a strong feeling we held that the supposed clients of research, in this case farmers, were rarely active participants in formulating the research questions.

Dusk was approaching as we returned from another absorbing interview. We were reflecting on what we had heard and our dissatisfaction with our 'team meeting' over lunch earlier in the day in which several team members had begun to develop a typology of farmers in terms of 'information rich' and 'information poor'. Descriptions such as 'information rich' and 'information poor', which were very much in vogue at the time, simply did not fit with our experience. There was a stereotyping of what constituted 'information' in the literature which was heavily biased towards whatever was technically 'the go' at any point in time. Our experience was that farmers we had interviewed were information-rich about those farming practices which they judged to be important to them . . . some of which were proving to be financially advantageous whilst others seemed to be breaking even or, occasionally, operating at a loss. Our experience was that farming families wanted to engage with us around those practices which most tapped into their energy, captured their imagination or represented what they most wanted to do. The generic term that best expressed this complex phenomenon was 'enthusiasm'.

The question we asked was, how much of the R&D that is being done leads to 'change that makes a difference'? The R&D literature is replete with examples of first-order change – change within the existing system, or more of the same; there appeared to us to be few cases of second-order change, change so fundamental, that the system itself is changed (Watzlawick, 1976; Russell and Ison, 1993). Watzlawick (1976) describes 'second-order change' as stepping outside the usual frame of reference and taking a meta-perspective.

This chapter relates our experiences of research designed to develop alternative models for future second-order R&D. It is also an exercise in reflection. Maturana and Varela (1987) describe reflection as a process of knowing what we know, an act of turning back upon ourselves. This process connects action and experience in a circular or recursive relationship such that 'every act of knowing brings forth a world . . . All doing is knowing and all knowing is doing.' (ibid p. 26).

6.2 Our contexts as researchers

How had we come to be together on the Wirrinya–Forbes road engrossed in a conversation in which the initial distictions about 'enthusiasm' arose? This is an important question about our context as researchers engaged in research. In the protocols that have evolved for publishing research little or no space is allocated to explore the context of the researchers, the research or the project. This may be because it is often perceived to be a messy state of affairs and outside the domain of research or science. We submit that it is an integral part of doing science or research and that critical reflections on project formulation and management warrant greater attention. Further, essential components of context, we argue, are statements about the researchers' epistemological framework.

Genuine collaboration, in our experience, does not just happen, nor is it readily forced upon people by way of grants and organisational structures. We suggest that genuine collaboration develops through relationships in which we are able to grow – to make new and satisfying meanings about our worlds of experience. We all bring to collaboration our histories of experience as well as our anticipated futures. This includes our theories about the world, our epistemologies, even if unconsciously. These shape the nature and creativity of the collaboration that ensues. For this reason we feel it important to relate our stories, as these provide the context for our collaboration in this research. We do so in a narrative style achieved by taping our reflections as the project was coming to a close.

David's story
After a few years at Hawkesbury (now the University of Western Sydney, Hawkesbury), with their system of education that encouraged people to design their own curriculum, I saw that for some people that worked very well and they jumped ahead in extraordinary leaps and bounds while for others it didn't seem to have any impact at all. In 1984–85 I came across the work of the Chilean Biologists, Maturana and Varela [1987]. It brought together my two interests – one was biology and the other was this thing about not being able to get inside other people's skin and influence them. We

can not pre-determine how someone is going to act – they will do what they want to do, and that will be connected to what you do, but you can not actually shape that person in any way. There it was, as plain as day – an explanation to what I had experienced myself. Not only did it make sense, but they said 'it is biologically so', and that made me feel good too. So it wasn't just a bit of philosophy, it was anchored to actual experiments with frogs and fish and pigeons. I liked the elegance of that experimental work – it looked very conclusive to me and I have never seen it challenged by anyone as being incorrect. So here was this beautiful explanation that I felt very comfortable with.

Then in 1989, Humberto Maturana offered a two-day seminar on his biology of cognition. Not only what he had written seemed to be just right, but here was this glorious person and he was uncompromising in what he said and how he said things. And he said it with such delight and excitement and fun. At the same time an American psychologist, Robert Johnson came out to Australia. He spoke for a couple of days, mainly through stories from all around the world. And he impressed me in the same way as Humberto did – here were two people who were excited about what they were doing, really believed in what they were doing, and had it beautifully connected. So while one was a biologist and the other a teller of mythological stories, I saw a strong connection between the two. I decided that it was quite compatible that when you are dealing with people and not salamanders, stories are a way of forming a connection. You are not trying to get into someone's skin and say 'to really get along in life you need to follow these principles and I'll give you a system of operating and monitoring them'. You just tell delightful stories and somehow that allows the other person to be connected to you but not feel imposed upon. This was the idea of narrative and I thought I could learn to tell stories just like them.

There is an historical tradition which leads to this idea of 'narrative'. In the period of the Enlightenment people believed they were discovering truth – getting closer to nature. So we saw laws of behaviour, laws of physics, laws of everything arising as a way of predicting what would happen in nature. Scientific method was the vehicle that came out of it, and a whole new language and culture arose around this search for truth.

Then came people like Hegel who had much earlier put forward the idea that truth cannot be discovered, it must be created. Not that any event or experience can legitimately be called a 'truth' but rather ... out of a particular relationship with its context, an idea takes on a personal meaning that can assume the standing of an 'experiential truth'. In much the same way, the telling of a personal story is akin to the expression of a personal truth. This perspective has been gaining momentum in all sorts of intellectual circles, *from psychology to literature and sociology. Writers who take a postmodern stance say that you can't have a data bank of truthful understanding – knowledge is created and the next generation will recreate it. From one generation to the next the metaphors change and we find the current metaphors more satisfying than the previous ones. So that draws a contrast between what we will call 'discovery' knowledge, and 'narrative' knowledge. Discovery knowledge presupposes there is truth in nature and we are continually getting closer and closer to a better understanding of it. Richard Rorty (1989) says that our culture is embedded in the 'facts of the matter' – our politicians are continually talking about it and so are people when they say 'it's common sense'. That is a way of our culture saying we are still very embedded in seeking out the truth, fundamentals, universal laws, facts.*

I ran a workshop for staff at Hawkesbury on these ideas, and it was Ray Ison and another colleague, Ian Valentine, who came up to me afterwards and said 'I like what you are saying, we should talk further about it.' That was very important to me at the time, because it made me feel valued. But then Ian went off to New Zealand and Ray went off to Sydney all at the same time and there didn't seem to be much hope of doing anything together.

Ray's story

An early experience with my father brought home to me some of the problems of the so-called technical expert. I had gone off to University enthusiastic about the role of pastures in improving farm productivity. With my new understandings I tried to persuade my father, a wool grower, to make changes in the way pastures were established and managed on the family farm. He was prepared to accept some of these ideas and he would put the initial steps into practice, but when it came to the next step he would often make what I judged to be the wrong decision if I were not there. It became clear that unless my father was able to take responsibility for developing his understanding of the whole issue and all of the subsequent steps that needed to be taken, then there was very little likelihood of long-term success.

So, very early on I drew back from attempting to be prescriptive, and telling people what to do. This matched up with my general attitude that the world was far too complex a place and that rarely were there single right answers to many of the issues we were confronting.

In 1978/79 I spent a year in Bali, Indonesia, doing field work for my PhD. Whilst my research was very traditional plot-based agronomy, the whole experience brought home to me the cultural dimension of agriculture. Agriculture in Bali was so much a process as opposed to what I'd been taught about as a more static, objectified agriculture in Australia. In Bali, religious ceremony, village life, family life are all intertwined around the agricultural practices and the rhythm of the seasons. I saw my own experiences from a new perspective. I was able to value my connectedness to place and appreciate more fully other aspects of my prior experience, such as my mother as farmer, something to which I was culturally blinkered in Australia.

Whilst in Bali I became interested in 'technology adoption' or how new innovations, particularly new pasture plants' might become incorporated into farming systems. It was clear that even very simple technologies would only become incorporated if people were able to encompass them within the cultural dimensions of their lives. Elephant grass [Pennisetum purpureum] was grown as a so-called improved pasture in Bali. It was very simply propagated, yet it had taken nearly thirty years to be widely used and distributed. I later learned and understood that the whole notion of growing pastures to feed animals was culturally alien and that when these ideas are totally outside people's experience, there is a need to think of other processes which occur when new technologies are introduced. The western models of technology transfer didn't seem to fit.

Later, I had the opportunity to do some consulting work for FAO [Food and Agriculture Organisation of the United Nations] in Tanzania. My experiences there had a most profound effect. It was a country on the verge of bankruptcy, yet a country that had immense potential, agriculturally and climatically. There were aid donors everywhere, all sorts of private agencies, and lots of development projects. In my work I became associated with a UNDP project based at Morogoro. They were attempting to develop State-owned ranches to increase the beef supply for the capital city. This project was typical of many UN projects in that the staff came from a great diversity of countries. The whole project to me seemed to be a fiasco. It became clear that each individual in the team brought so much cultural baggage to Tanzania and so many expectations of how things might be or could be. Because the westerners had the money, they set up the systems, pushed things along, and as soon as they pulled out, the whole thing collapsed around them. I saw that the only form of sustainable development was where those who were part of the country or the industry were the ones who took responsibility for doing it themselves.

I moved to Hawkesbury in 1982 and the innovative educational system was something that I related to very readily given my previous experiences (Bawden et al., 1985). I was particularly keen to take some of the ideas that were being formulated for the curriculum beyond the institution and to work in the broader community (e.g. Ison, Potts and Beale, 1989). At about this time I also went to a seminar presented by David after he had returned *from study leave. He had written a monograph which I found very stimulating and I remember talking to him about it after the seminar and being keen to talk further. Shortly after I left Hawkesbury to work at the University of Sydney.*

At Sydney I became increasingly concerned with what I call the 'problem metaphor'; my concern was to explore the problem formulation process, initially for agronomic, but subsequently for agricultural R&D. This seemed to be the neglected critical first step in the project cycle – how were problems formulated and who was involved in this process? In late 1986 I and Peter Ampt, a tutor interested in doing postgraduate study, began exploring the potential and utility of Rapid Rural Appraisal or RRA, a qualitative survey methodology, in a so-called 'developed' country setting. RRA was then increasingly being used in third world countries to better formulate problems for research. The idea of a RRA was something we agreed was of interest and it led to the design and conduct of the RRA in the Forbes Shire in central western New South Wales in 1987/88 (Ison and Ampt 1992). In forming the multidisciplinary team we invited a number of people including David to be involved. We were particularly concerned to develop good group function as a means of breaking down strict disciplinary boundaries and David was invited to participate and to take a role as participant observer so as to evaluate our group process.

It is from these histories that our current collaboration arose in the period 1990–1993.

6.3 Describing the phenomenon we had experienced

As seems often to be the case with research, projects arise because of particular histories of the researchers. The project around which this book, and more specifically, this chapter, is based had its origins in our conversation on the Wirrinya to Forbes road.

In the final stages of the RRA, in 1988, in which the phrases 'information rich' and 'information poor' were being used by RRA team members to describe categories of farmers, the team split into three sub-groups to prepare preliminary reports around issues which each group felt captured the most important outcomes. In our group we began to question the adequacy of the language, underlying concepts and thus categorisation of producers in these terms. We shared a dissatisfaction with these terms as organising metaphors – they did not coherently synthesise our experience. David had been talking about the idea of stories and people having different levels of energy to do different things. We were struggling with the distinction between information-rich and information-poor when we realised that the notion of 'enthusiasm' we had been talking about earlier was a way of understanding what was happening. Farmers really did seem to make sense of all sorts of complexities and do all sorts of interesting things when they were enthusiastic about it. It rang true for us. There was a great deal of excitement at the time and we sat down to talk about enthusiasm and how it might be better understood and used as a basis for R&D. David had initially been inspired by the notion of enthusiasm proposed by Robert Johnson (1987) in his charming little book *Ecstasy. Understanding the Psychology of Joy*.

Initially we saw *enthusiasm* as a higher order concept in which those who were 'enthusiastic' appeared more able to manage their own realities and to accommodate change. Thus our experience with the limitations of 'information rich and poor' and our conscious introduction of the alternative metaphor 'enthusiasm' can be seen as a critical incident. In describing our research we draw heavily on what we term critical incidents. A critical incident is where an existing organising metaphor was unable to coherently organise or synthesise our experience. We thus consciously introduced one which we felt did the job better.

From this experience it was proposed that it might be possible to generate or trigger enthusiasm in people by valuing and appreciating them for who they were and for what they were doing, not unlike we had been doing in our very open-ended interviews with farmers. Our proposition assumed all people were information rich and thus capable of making the best sense of their realities. Based on our experience we proposed that a key feature of triggering enthusiasm would be the setting up of processes so that people were listened to with acceptance for what they had done, for what they were doing and for how they thought. The concept of 'triggering' is critical here as we proposed that enthusiasm was already present and was not something that could be instilled by another. The assumption was that everyone is enthusiastic about certain key features of what they do and/or what they believe needs to be done. They are not enthusiastic about everything they do even though they might perform effectively across a wide range of practices. What became obvious to us as researchers was that each individual possessed a reservoir of unexpended energy or excitement which was a resource for collective action if it could be effectively organised (elicited and brought together) in some manner. We saw the opportunity of developing a process to provide a mechanism for the expression or development of enthusiasm as a research question.

There were other events which encouraged us to push on. One indirect consequence of our RRA collaboration was to further collaborate on a review of the conceptual basis of rural extension theory and practice (Russell *et al.*,

1989). Further experiences during the RRA increased our connection with this issue. Our experience was that many technologies that had been developed by researchers were not being used; this added weight to the argument that the traditional research–extension–producer model for the transfer of technology had not been successful in achieving its aims (this is described more fully in Part I and by Russell and Ison, 1993; Fell and Russell, 1993).

6.4 Enthusiasm as theory, driving force and methodology: proposing an explanation for this phenomenon

In thinking about our research approach we wanted to incorporate the idea that enthusiasm is several things. It is an intellectual notion or a theory, it is an emotional thing or a driving force (not necessarily connected to any reality-testing process), and it is a methodology of how to do (an observable strategy to go from A to B). We wished to do all three as best we could.

As an intellectual or theoretical notion, the original meaning of the word enthusiasm goes back to the Greek words '*en*', meaning within, and '*theos*' meaning god. So the word captures the notion of the 'god within' as distinct from the source of all understanding being from without. Throughout Western History there has been tension between whether the primary focus of our understanding comes from nature or from within ourselves (Johnson, 1987).

The emotion or driving force idea of enthusiasm has always been central in psychology. Motivation has been understood as a drive from within that then gets satisfied by whatever you are doing outside. And so the drive of hunger gets satisfied when you eat and the drive of sex is satisfied when you have sex. It comes from within and remains unquelled until it is satisfied and then it calms and rises up again. In motivation theory it is always originated as a biological drive. Enthusiasm could then be conceived of as a drive much like hunger in as much as it is the drive to do. But it is not targeted like hunger. It is the thing that gets us up in the morning to face the world. It acts as a source of meaning which provides the energy that helps us do what we want to do. It gets satisfied in a similar way to the biological drive when we find ourself doing what we wanted to do. We feel the same satisfaction as we do when we are having a meal – satisfying the drive of hunger. And then like any other drive it then quells and returns. So enthusiasm fitted the general notion of a drive in a biological sense – the engine of life. This is further exemplified by Tucker (1972, p. 160) citing Melmoth who thought enthusiasm all to the good, and spoke of its forces as 'some of the main wheels of society' and 'indeed we should agree that it usually is what makes things go'.

Enthusiasm as a methodology then is where the use of narrative comes in. The methodology must be underpinned by the biological understanding of the drive itself as well as the theoretical principles. It must be shaped in a way that does not re-direct a person's energy – we want to find out where their energy is. This is the initial task in the methodology. But to do that we need the right sort of questions: what do you want to do, why are you a grazier, what is it you get out of this sort of work that is satisfying? To get to the point of having that sort of conversation requires a respect for the individuality of the other and acceptance that whatever they are going to say is valid – based on the notion that it is the god within that person which has to be respected – that is where their energy is. So they can tell their story about where their energy comes from and how they see it expressing itself and what they see as obstacles to its manifestation.

We strongly believed that the only way you could gain these sort of stories was to cultivate the ground to get narrative. Asking questions like 'How many hectares do you have?' and 'How many sheep do you run per hectare?' can only elicit facts. These sort, of questions come from the scientific culture which believes that there is truth in nature and we should be continually trying to get closer to a better understanding of it. Your worth as a person is judged on the amount of information you know. We were not about that sort of knowledge. To get a story, you need to ask a different kind of question, and use a different kind of language. So rather than asking 'how many...' we needed to ask 'how did you come to be here?', 'what keeps you here?', 'why do you want your children here?' These are questions which allow a person to create a story, not relate facts. So to incorporate the two ideas, by encouraging someone to tell a story, you are getting a glimpse of what drives them – of the metaphors that mean something to them. You don't get that with discovery knowledge. You get a sense of power by feeling that 'my ideas are better than yours' because there is a continual competition around who is closest to the truth, who has the greater body of knowledge, who has the most facts. Narrative and enthusiasm are independent but related concepts. They hang together because one expresses the other. Narrative allows us to find out where people's enthusiasms lie. It therefore allows us to develop a methodology around the concept of enthusiasm so we can actually go out and talk to people about their enthusiasm by asking them to tell stories about their lives.

By listening to and encouraging these stories, we hoped to be able to recognise moments of enthusiasm. This would be when the person showed biological emotion – their body changes, their voice changes, and they become excited. This seemed to be inhibited by matter-of-fact questions. The enthusiasms would be brought together and mapped out to be presented back to the graziers. Then we could invite them to work together on their common enthusiasms. In this we would play a supporting role and also observe the process.

The final aspect of the methodology is that when there is a climate established of people being encouraged to express their enthusiasms that this will be a mutually satisfying environment. If people share a common enthusiasm then they should work very well together. All three ideas of enthusiasm together gave us the direction for the project.

6.5 The research project: deducing from the first experience other experiences which were coherent

In late 1989 an opportunity arose to undertake a research project in which our notions of enthusiasm could be further developed. Australia has many rural industry research corporations, or RIRCs as they are commonly called. These bodies are funded by levies on producers with matching grants from the federal government. We had been involved previously with the Wool Research and Development Corporation (WRDC) who had commissioned us to undertake the review of rural extension theory and practice. The outcomes of this review convinced the then somewhat sceptical board of WRDC that it should become involved in funding 'technology transfer research'. At this time there was a strong sense of concern in certain circles that many of the technologies developed by researchers for the sheep-based pastoral industries of semi-arid southern Australia, were not being adopted. This was discussed more fully in Chapters 3 and 5. We were subsequently commissioned to undertake a research project to address these concerns.

The stated project objectives based on the project proposal to the WRDC were to:

1. Identify the processes used by Western Division (the area of semi-arid pastoral land in Western NSW) sheep graziers to secure and evaluate information and the factors influencing their use or rejection of technology.
2. Develop more integrative and participatory technology transfer systems based on the results generated from (1)
3. Test these systems for adoption/adaptation locally and beyond.

The language used to define R&D objectives of this type is not capable of revealing the nature and quality of the relationships of those involved in their formulation nor of the varying conceptions of meaning attached to the terms by the various participants. For example the principal researchers did not hold the common conceptions of 'extension' and 'technology transfer' then evident in the agricultural R&D community (Russell *et al.*, 1989; Ison, 1990). It was, however, felt necessary to draw on terms in common usage in the project literature. Later in the research team-building process it was felt necessary to explore these differences. The concepts which initially shaped the project came from a discursive field (Foucault, 1972) which included: rural extension (Russell et al., 1989); social ecology (Russell, 1986, 1992); systems agriculture (Bawden et al., 1989) search conferencing (Emery et al., 1978); participatory rural appraisal (Ison and Ampt, 1992) and the biology of cognition and perception (Maturana and Varela, 1987). The project, initially designed as a three-year collaborative study, involved three principle researchers: one each from the University of Sydney, the University of Western Sydney (Hawkesbury), and NSW Agriculture, the state-based agricultural R&D organisation.[23]

We drew on enthusiasm as a theoretical principle to guide the design of our:

 (i) initial research team-building workshop;
 (ii) research with graziers which included the design of a process for a community meeting (Chapter 7), the facilitation of grazier R&D action, and evaluation;
 (iii) research with rangeland R&D service personnel (CARR, 1993a) and
 (iv) own ways of working together as a team.

In the remainder of this chapter we wish to report critical incidents, and organising concepts, which elucidate the phenomenon of enthusiasm. It is not possible to detail all of our research here and readers are referred to subsequent chapters for more detail.

6.5.1 *Team building and the research design*

We had known from our earlier experiences that collaborative, team-based research was often difficult. Many researchers have large egos and are often committed to their particular way of seeing the world and doing things. We were aware that it was easy to say we were involved in, or to write about, multidisciplinary or interdisciplinary research, but what did this mean in practical and theoretical terms? Thus when our team came together we felt it important to address these issues. Our concept of enthusiasm guided our design of the subsequent group process.

In the initial team meeting each principal researcher narrated their personal histories and current enthusiasms to the group as well as their story of the project origins and their role in its development. A researcher from NSW

---

23 / The principal researcher from NSW Agriculture withdrew from the project prior to its completion so as to take up the position of Western Lands Commissioner, responsible for the administration of the Western Division of NSW.

Agriculture who was also on the WRDC and who had participated in the meeting where funding was agreed, was also invited to participate in this process. Semi-structured interviews (SSIs) were used in the workshop as a means to appreciate and value each other's potential to contribute to the group – to appreciate our diversity. This enabled all participants to experience both aspects of the interview process and to simulate what was proposed for our field work: the mapping out of individual's patterns of meaning over time – first the historical context, second the present-to-hand experience and third the anticipated context (the future).

The stories which emerged provided a very rich picture of the project origins and helped to place it in a context – different, original, creative and important to a range of stakeholders. It also provided a strong basis for relationship building between members of the research team, and to some extent with the principal funding body. This collaboration at the 'coalface' enhanced understanding of what each expected from the other, and most importantly, what the team was planning to do. The process focused on the use of narrative; individuals were listened to unconditionally and not judged with respect to personal value systems.

6.5.2 Semi-structured interviews with graziers

As researchers we were aware that we had little appreciation of the context of graziers in the semi-arid rangelands of Western New South Wales. As a means to better appreciate graziers' context, the team divided into three groups and conducted a series of in-depth conversations with a minimum of three graziers and their families across three north–south transects (Conway, 1985) of c. 240 km. On the basis of these transects the initial research with graziers was located 110 km north of the city of Broken Hill and centred on a region called Fowler's Gap (see Chapter 7 for more detail).

Sixteen pastoralist families (a sample of 48%, chosen at random) were invited to tell of their day-to-day experience in interviews conducted in their homes. In our initial contact we were clear that *we* were seeking *their* assistance with the conduct of *our* research, which was to attempt to develop more effective R&D systems. For the majority of graziers it was their first contact with anyone from the WRDC. The aim was not to pursue a 'fact finding' mission, but rather by the medium of stories, for them to tell of their experience. The interviews were used to trigger these accounts and were designed to map out patterns of meaning over time.

At the completion of the interviews our concept of enthusiasm was used to prepare for a meeting to which all of the graziers and their families in the chosen region around Fowler's Gap were invited. The major objective of the workshop was to invite graziers to become researchers on an issue or issues of concern to them. The full workshop process that was developed is described in Chapter 7. Of importance here are:

(i) each interviewer was asked to reflect on their experience of the interviews and to prepare a poster on a theme that they as researchers were most enthusiastic about. These posters were used during the subsequent meeting to 'mirror back' the understandings we, as interviewers, had gained from the process. We attempted to be clear at all times that these were our interpretations of what we had heard during the interviews and that the themes chosen were ones that had excited us, they were not accounts of 'how things really were'. The metaphor of 'mirroring' was incorporated in our design of process as an attempt to 'trigger' new ways for graziers to see their context;

(ii) it was also necessary to explore with the graziers interpretations of the word 'research' – most graziers in the region had little to no direct contact with researchers and pictured research as something abstract, done in laboratories or on research stations and isolated from their experience. It was found useful to distinguish between (a) research on things or plants and animals; (b) research on people and (c) research with people. Graziers found these distinctions meaningful, and were intrigued by our inviting them to do the latter, particularly given that some had experienced researchers as taking and never seeming to give anything back in return.

(iii) 'mirroring' and the deconstruction of 'research' were used as precursors to inviting graziers to discuss and articulate in facilitated small groups the issues on which they were particularly keen to take research action. This aspect of our process design was in reaction to earlier experiences with participatory research approaches in which local people were often left out of the process of making sense of data and of the ownership which comes from this important meaning-making stage;

(iv) each individual was given the opportunity to contribute to the discussion in the small groups following which each individual nominated on paper the issues they were most keen to research. From the first meeting four sets of issues emerged:
(a) the marketing of middle micron wool;
(b) secondary and tertiary education for children;
(c) community development and
(d) interest rates.

(v) Participants were then invited to form groups around the issue they were most keen to take action on. Of those present the largest group were enthusiastic about a project on 'the marketing of middle micron wool'. We agreed to return after inviting the graziers present to express whether they wished us to come back to help them with the research they had nominated and after agreeing that it would be desirable to repeat the workshop with a bigger group. The graziers expressed their desire to contact their friends and neighbours and to invite them along.

(vi) As these pastoralists shared a common geographical area, it was thought possible to form groupings around shared enthusiasms. Our process resulted in graziers attending who were not normally involved in local organisations. Apart from one family, all those attending were from the 16 families interviewed. Those participating extended invitations to their neighbours for the next meeting but attendance did not increase. Light falls of rain precluded travel of some graziers on the occasion of both meetings. In the initial workshop four 'enthusiasms for action were nominated'. By the time we returned for the next workshop the group presented us with a 'consensus' decision that they wished to research 'the marketing of middle micron wool'. Upon reflection this represented a critical incident in the research process. It was at this stage that the process we had devised, and which was taken on by some of the graziers in the second workshop, moved away from the individual expressions of enthusiasm. The implications of this are discussed below.

(vii) The project to 'investigate the marketing of middle micron wool' continued over a two-year period and is reported in a document developed collaboratively with those participating (CARR, 1993b). All research action was nominated by graziers and facilitated by a member of the research team. Graziers who participated in this research describe their experiences and the outcomes for them in Chapter 8.

(viii) Follow-up conversations – some twelve months after our initial round of interviews there remained tension and concern amongst our research team and participating graziers about the number of graziers actively participating in the project. We had attempted to keep in touch with all local residents by writing a personalised newsletter, developed collaboratively with participating graziers, which reported on events and which always reissued the invitation to be involved. At this point we made a round of visits to graziers in the region with the purpose of inviting them to join in the '*network of conversation*' that constituted the project. We felt that this might be achieved through the further development of personal relationships and, by active listening, giving space to the other. It became apparent that most were aware of the project. We attempted to be careful not to have the exercise seen as a means of finding out why people had not participated: conversations were held with graziers who had been actively involved as well as with those who had not. Two major themes emerged from these conversations: (i) the combination of severe drought and low wool prices were causing severe financial and emotional pressure on most families. Some families responded to this situation by becoming more isolated and withdrawn whilst some others saw it as part of the vagaries of their environment and that all one had to do 'was shut up shop' and wait till it passed. Obviously the latter strategy was not available to those graziers with low equity or high interest repayments; (ii) the round of conversations crystallised earlier experiences to allow us to articulate another critical incident. Most graziers rarely talked in any detail about their management practices with their neighbours. However when invited they were always enthusiastic to do so in considerable detail. This resonated with our own experience both in our early life in rural Australia and in our current relationships with our partners. We proposed an explanation that in certain sections of Australian society there exists a way of living which deems it inappropriate to impose your ideas on to other people or to be seen to be 'bragging'. The converse of this was that it is also seen as inappropriate to be a 'sticky beak'[24] and to want to know too much of your neighbours affairs. This insight led us to design a process of sharing of experiences through public semi-structured interviews (see below). It also provided insights into some of the problems with the current research – extension – client relationship, namely that rarely were genuine invitations made to graziers to participate and to share their understandings with researchers. As Humberto Maturana (pers. comm., 1993) has observed 'for an invitation to be a "true" invitation it needs to be equally

24 / Australian colloquial term – to interfere in someone else's business.

acceptable if it is "accepted" or "rejected". If it happens to be turned down and you, the one who offered the invitation, become angry and feel rejected, then you must conclude that it was never an invitation in the true sense of the word'.

(ix) Sharing research action and critical evaluation: planning and critical evaluation of all the actions, in a community setting, made this research as opposed to indiscriminate action. Graziers agreed that their research was beneficial, and that some graziers had 'got a lot out of it' (see Chapter 8). Graziers had received insights into what happened when the wool left the property, and some changes in on- and off-farm practices resulted. Some graziers were constrained, both socially and financially, to continue present marketing practices.

Graziers also began to see themselves in a different light. They appreciated through this experience that they were also researchers with something to contribute to the R&D process. There remains within that community scope to continue research into this issue, or indeed to switch to issues that become relevant. This sense that research can be 'on-going' was exemplified by a further activity which was arranged by graziers and took place after the official end of fieldwork by the project team.

Table 6.1 details a schema developed to evaluate the research conducted by the graziers covering past, present and future concerns, actions and usefulness. This was completed during a community meeting in February 1992, when individual graziers were invited to tell of their research actions in a public semi-structured interview.

6.5.3 Working together as a team

In our team process and management we attempted to value and listen to each person's enthusiasms for action. This did not always work as some members of the team demanded more time for their own issues than it proved possible to give. This was addressed by recognising the different roles and responsibilities of principal researchers as opposed to research assistants. Meeting rituals were also developed to facilitate emotional as well as intellectual appreciation. Meetings commenced with a 'tea ceremony' in which food and drink were shared and each team member was invited to relate a story which reflected their situation since the last meeting. Participants were also invited to name and list the anxieties and/or excitements they had brought to the meeting. All meetings concluded with each person providing their assessment of the day, clarifying expectations, and nominating issues for future agendas. By doing this we were trying to develop what

Table 6.1
A schema used for the evaluation of the collaborative research project formed around the common enthusiasm of marketing of middle micron wool

	Past	Present	Future
Driving force/ concerns	*Lack of AWC[a] commitment to 'Bread and Butter' wool promotion *Not knowing what promotion was happening *Seeing how our money is spent *Getting the shearers to do the right job	*AWC still don't understand our concerns *'Elite' promotion perspective *Ensure that this (our R&D) keeps happening	To do a better job at the micron that you're doing What we want needs to be better 'transferred' to the AWC CSIRO[b] Melbourne (new processing technology)
Actions	Visits: Shearers to Michell's[c] Growers to Michell's Growers to AWC Growers to wool mill in Wagga Grower to IWS[d] Information gathering: marketing; promotion; micron; clip preparation Experimentation: micron change; direct selling	Sales of wool to alternative buyers Instil more pride in shearer's work Better sheep presentation in shed	Going back to AWC Working to greater uniformity of clip AWC has picked up its game re 'our wools' Stud-based promotion/sales
Usefulness		Direct feedback on clip from Michell's Direct selling is a viable option No reason to change our type of wool Better base for decision making More confidence to 'find out' (research) Most research done on the wrong scale	A more even and stronger fibre Our own experimentation doesn't need to be accelerated (e.g. stocking, water, micron, wool type)

[a] AWC, Australian Wool Corporation.
[b] CSIRO, Commonwealth Scientific and Industrial Research Organisation.
[c] Michell's is a private buying and processing firm; traditionally most graziers have sold through the auction system and not to private firms.
[d] IWS, International Wool Secretariate.

Robert Johnson (1987) describes as meaningful ritual and ceremony to 'find our way between society's expectations and our spiritual needs' which is 'the way that makes the impossible possible'. Ritual is also seen by Johnson as a means to foster enthusiasm.

6.6 Experiencing once again the phenomenon of enthusiasm

The test we had set ourselves in the design of the research was to see whether, through our joint activities, and our attempts to have our theory guide our design of process, we as researchers could observe once again phenomena that were coherent with our initial experience of enthusiasm in the Forbes RRA. We felt this was necessary to be able to give a technical understanding of enthusiasm and to complete the research cycle.

The experience of our project has led us to define the triggering of enthusiasm for action as a four-stage process:

(i) When one is invited to talk (to tell of their experience past, present and anticipated future), the emotional connection occurs through active listening – 'I really want to hear what you have got to say'. Genuine concern for the other is manifest in the conversation (language in all its manifestations). Practices to create the environment for enthusiasm to emerge through narrative are summarised in Box 6.1.

(ii) In the course of being listened to (showing respect, not prejudging, nor pushing one's own agenda) an invitation is made consciously or unconsciously. This provides space for options which lead to mutually satisfying action to be generated. In this process one is not naming these options for the other. In this form of interaction there is the possibility of self-awareness and triggering of latent ideas or concepts.

(iii) Action is taken in the domain of interactions – to maintain the conversation. Processes (i) and (ii) will not lead to all individuals joining the conversation. Action does and continues to occur in other domains. For some, however, new ways of being are triggered, the possibilities for which existed all the time. Thus there is no transfer of anything nor working towards predetermined plans or goals. What is triggered is what Maturana (1988, p. 42) describes as emotions and moods or body dispositions for actions (he distinguishes 'moods as emotions in which the observer does not distinguish directionality or possibility of an end for the type of actions that he or she expects the other to perform').

(iv) External resources (e.g. money, technology) become amplifiers or suppressors of enthusiasms for action. In our research we have encountered both.

A further two stages are necessary for the development of enthusiasm as R&D methodology:

(v) Careful attention is necessary in the design of processes to bring people together who share common enthusiasms for action. Reflecting on our own experience we identified a critical incident relating to the group consensus that was presented to us at the second meeting (described above). From this experience and the subsequent lack of action by those who were not initially enthusiastic about the marketing of middle micron wool, we would propose that consensus suppresses enthusiasm for action and that at many levels of system activity the search for consensus is an inappropriate objective (the tyranny of agreement). We experience this in everyday life when in our relationships we are often forced to compromise and lose our energy and vitality for action. The opposite to consensus is to value diversity of both research action and of modes of 'being a pastoralist' (we return to this theme in Chapter 9).

(vi) Prior to about 1829 the word 'enthusiasm' was associated with the emergence of the new radical religions and was seen as an emotional driving power devoid of reason and rationality. As the discourses of the enlightenment gained momentum the meaning accorded enthusiasm exemplified the tensions that continue to exist between the first- and second-order traditions we describe in Chapter 1. The choice was '...between enthusiasm and the rational worship of God' and for some it was seen as the 'offspring of passion and disordered intellect' (Tucker, 1972). To avoid the prospect of enthusiasm giving rise to disordered intellect and action, we propose the need for cycles of critical reflection to be an essential part of the use of enthusiasm as methodology. There are no doubt many ways which this could be achieved. Our Table 6.1 provides one example which was judged to be useful by all involved in our project. Box 6.2 lists some of the principles we would now adopt to guide us in the design of processes for participative R&D utilising enthusiasm as methodology.

One of the times of greatest enthusiasm for us was following our initial conversations with graziers (transects), but this gave rise to some difficulties

> **Box 6.1** The theoretical insights generated in the research were used to *inform* the development of *guiding principles* for process design involving enthusiasm as methodology. These were refined on reflection on our experiences working with graziers and service personnel:
>
> - People value being invited to contribute, even if the invitation is not taken up. It is desirable for invitations to remain open for all to participate in the research if they want to.
> - Each person's knowledge and experience is unique and valid for its context. Knowledge and experience are results of the particular context (personal and physical environment), and specific to individuals and their social situation.
> - The context in which knowledge and experience are gained is very important and often overlooked.
> - Research is not just a 'Scientific Approach'. Graziers were engaged in valid research, as were many extension officers although they and their 'researcher' colleagues were reluctant to acknowledge this. A diversity of research approaches is valuable as such diversity is more likely to underscore the context-dependent nature of research.
> - Research which links theory with action through a process of planning, acting, observing and reflecting or evaluating is likely to lead to more meaningful outcomes than just discovering the 'facts'.
> - Each meeting needs to be carefully designed with attention to process, use of language and sharing of expectations.
> - Research often does not work if someone else always takes responsibility for every aspect of it. If this happens local people or the 'clients' have no power and responsibility. This may mean that research is not useful in many situations.

in research design as a compromise had to be made as to which grazier group to invite to work with us. It had been our intention to invite sequentially all groups from the initial conversations to join in the research. This did not eventuate and gave rise to a parallel situation when graziers adopted a consensus approach; that is there was a lower level of emotional commitment to the conduct and design of the grazier-related research by some team members than might otherwise have been the case. An interpretation of what happened, based on Maturana's (1988) position, was that some of the research team were invited to enter a conversation in which they had no emotional commitment at a time when they were enthusiastic about their own community, realised through the conversations of the transects. As a result of the decision to work at Fowlers Gap and the absence of the main researcher employed to do the field research from the SSIs, the emotional connection via enthusiasm did not eventuate for this researcher until the round of conversations conducted in late 1991. As an explanation, we would propose that in accepting the graziers' consensus decision, which conveniently met our needs, and in geographically concentrating our research effort following the transects, we were 'seduced' from the integrity of our theoretical position.

> **Box 6.2 Practice designed to create the environment for triggering enthusiasm**
>
> 1. Let go your own thoughts and awareness so as to give your full attention to the other person.
> 2. Show in your initial exchange that you really want to hear what the other person is saying and in the way they are saying it.
> 3. Encourage a climate in which others in the room accept that what is being expressed is valid for the speaker and can be understood as such without demanding 'agreement'.
> 4. The facilitator (trigger person) needs to manage the environment so that each person feels that they will be listened to equally and what they say will be accepted (not necessarily agreed with).
> 5. Encourage each person to say more by having 'an ear' for the plot. Encourage the person to move through hesitancy.
> 6. Recognise that some leads will in turn trigger a more generalised enthusiasm. A particular story may unfold (gain public recognition) at another time and place.
> 7. The trigger person must show sensitivity at all times to the story as it is told. Treat the story as a precious gift. This can be shown in verbal and non-verbal behaviour.
> 8. Be aware that there can be interrupting behaviours, passive, negative, dominating, etc.

Tucker (1972) also draws attention to the transience of enthusiasm. This raises the question of how enthusiasm, the emotional state, is nurtured for meaningful and desirable action. In other words how might cycles of enthusiasm, action and critical reflection be triggered, and sustained in a recursive loop within a consensual domain that also has some geographical character? We respond to this question in Chapter 9 but it is clearly a question for further research.

6.7 Reflections and future directions

Maturana and Varela (1987) and Maturana (1988), based on their neurobiological research and their concern for the question 'how is it that we can know?' have proposed that doing science, or providing a scientific explanation for a phenomenon, can be best described as:

(i) *describing* a phenomenon that has been experienced and doing this in a way that allows others to agree or disagree as to its existence;

(ii) *proposing* an explanation for the existence of this phenomenon. This explanation functions as a 'generative mechanism' in the sense that when the mechanism operates the phenomenon appears;

(iii) *deducing* from the first experience other experiences that are coherent with the first, and which would result from the operation of this mechanism that has been proposed as an explanation; and

(iv) *experiencing* the other phenomena that were deduced in step 3. While quantification is not essential to this process it is often useful in the deductive phase.

Thus experience is explained with an experience and the generated explanation always remains secondary to the world of daily living. The research project we have described is our attempt to do science from this perspective. We readily admit to having no prior experience in doing science from this perspective and we know of no others who have attempted to apply this model of doing research in this domain.

Maturana further suggests that what we accept as an explanation has both a generative mechanism (the operation of enthusiasm in this case) and an informal condition that must be satisfied (this is elaborated on in the Appendix). When reviewing an earlier version of this chapter a journal editor expressed his scepticism of this version of doing science. One is tempted to suggest that our explanation did not meet his informal conditions for what it is to do science. We do not say this to be critical of this most constructive reviewer, but merely to point out that 'informal conditions' such as exist in our culture of what it is to do science, can be very powerful. The appeal of this model for us is that it grounds us in experience, and our experience is both a biological and social phenomenon. This is a paradoxical situation. As Gareth Morgan (1993) observes: 'while humans can . . . be seen as active agents in perceiving, constructing and acting on their worlds, they do so in circumstances that are not their own choosing' because 'as philosophers like Michel Foucault[25] have shown, there are all types of power relations embedded in the language, routines and discourses that shape everyday life' ... 'which lock people into feeling that they are hemmed in by deterministic forces over which they have no control'. Of course doing science is part of everyday life for a scientist who lives in the discourses of science.

Reflecting on our initial experience in which the metaphors of 'information rich' and 'information poor' were inadequate reveals to us two things: (i) just how powerful and obfuscating accepted metaphors or discourses can be; it takes very powerful experiences and human support (of some kind) to break free. By break free we mean to bring oneself to the position of engaging in critical reflection; (ii) there are many examples from everyday life of people coming together around common enthusiasms for action. We are sure that you have experienced these, mostly positively, but sometimes

25 / The ideas of Foucault were taken up by Adrian Mackenzie in Chapter 4.

negatively as history can show us. This suggests that not only should it be possible to investigate the experience of enthusiasm in the mode of scientific inquiry we have taken from Maturana (1988) but also empirically by seeking out groups which appear to have formed around some common enthusiasms. Such an approach may reveal how enthusiasms have been triggered and whether something an observer could name as critical reflection has been built into their group function. Epistemologically, however, this would be in the domain of first-order R&D – where the espoused stance of the observer is outside the system under study, which is different from the research we have conducted. We explored this in more detail in Chapter 1 in which we reviewed the contemporary context of R&D in rangeland settings. Our approach, or second-order R&D, where the researcher is part of the system under study, differs in that it reveals the emotional connections between the researchers and, in our case, the attempts to experience the interplay of the triggering of the emotion of enthusiasm, the subsequent action and evaluation.

Our experience strongly suggests that traditional disciplinary expertise and the practices that have become institutionalised in so-called 'service agencies' cuts across local capacity to coalesce around common enthusiasms, often undermines it and in some instances appropriates local initiatives to meet institutional rather than local needs (see Chapter 5 and CARR, 1993a). Tucker (1972, p. 112) observed that: 'Science as well as love or religion could lead its enthusiasts into unsociable behaviour: it inspires people with so fond a passion for particular studies that by degrees they acquire a dislike and frequently a contempt for all objects which do not coincide with their own pursuits'. In this statement can be seen the concerns of many about the tyrannies of the academic disciplines. John Cobb (1993) has described this as 'disciplinolotry' and like 'all idolatries' he is opposed to it. This has deeper roots in his concern for the organisation of knowledge in the modern world and the modern university, which he sees as the embodiment of this discourse.

As Maturana points out, there are the conversations of work and the conversations of mutual acceptance; when a grazier and researcher carry out their conversation in a manner that entails mutual acceptance, then such a phenomenon is founded on a biological state that might be called the emotion of respect, specifically respect for human differences. Maturana refers to this emotion as 'love' and denotes any conversation based on this emotion as constituting what is essential for a sustainable society. He claims that interactions that do not entail mutual acceptance between people are most often recognisable as interactions (or systems) which entail one person trying to dominate another. The same applies for researchers from different disciplines and is at the core of what we seek, but rarely find, when we use the word 'interdisciplinary' to describe a form of research. This is a theme we return to in Chapter 9, when new forms for organising R&D are explored.

Acknowledgments

The authors would like to thank Dennis Gamble, David McClintock, Susan Wyndham, Lynn Webber, Craig Pearson and Cathy Humphreys for assistance in preparing this chapter. We also acknowledge the support and participation of our co-researchers.

References

Bawden, R.J., Hannibal, M. and Gahan, C. (1989). On Knowledge, Knowing and Change: Farmers in Semi-Arid NSW. Unpublished Manuscript, University of Western Sydney (Hawkesbury). Richmond.

Bawden, R.J., Ison, R.L., Macadam, R.D., Packham, R.G. and Valentine, I. (1985). A research paradigm for systems agriculture. In *Farming Systems Research. Australian Expertise for Third World Agriculture*, ed. J.G. Remenyi, pp. 31–42. ACIAR, Canberra.

Bawden, R.J. and Ison, R.L. (1992). The purpose of field-crop ecosystems: social and economic aspects. In *Field-Crop Ecosystems*, ed. C.J. Pearson, pp. 11–35. Elsevier, Amsterdam.

CARR (Community Approaches to Rangelands Research) (1993a). *Institutions Responding to a Community Request: Wool Marketing Advisory Support Group ... and Beyond*. Monograph. University of Sydney and University of Western Sydney. 48 pp.

CARR (Community Approaches to Rangelands Research) (1993b). *Marketing of Middle Micron Wool. Research with People on Issues that Make a Difference*. Monograph. University of Sydney and University of Western Sydney. 44 pp.

Cobb, J.B. (1993). The Cosmos and God: The Dependence of Science on Faith. The 1993 Templeton Lecture, Centre for Human Aspects of Science and Technology, University of Sydney.

Conway, G.R. (1985). Agro-ecosystems Analysis. *Agricultural Administration*, **20**, 31–55.

Emery, F., Bucklow, M., Foster, M., Thorsrud, E., Trist, E.L., Wilson, A.T.M. and Woolard, W. (1978). *The Emergence of a New Paradigm for Work*. Centre for Continuing Education, Australian National University, Canberra.

Fell, L. (1993). Epistemology and feedlot cattle. Unpublished manuscript, Elizabeth Macarthur Research Institute, Camden, NSW.

Fell, L. and Russell, D.B. (1993). Co-drifting: the biology of living together. (Unpublished manuscript.)

Foucault, M. (1972). *Archaeology of Knowledge and the Discourse of Language*. Pantheon, New York.

Ison, R.L. (1990). *Teaching Threatens Sustainable Agriculture*. Gatekeeper Series No. 21. International Institute for Environment and Development, Sustainable Agriculture Program, London.

Ison, R.L. and Ampt, P.R. (1992). Rapid rural appraisal: a participatory problem formulation method relevant to Australian agriculture. *Agricultural Systems*, **38**, 363–86.

Ison, R.L., Potts, W.H.C. and Beale, G. (1989). Improving herbage seed industry productivity and stability through action research. *Proceedings of the XVI International Grassland Congress, Nice*, pp. 685–86. Association Francaise pour la Production Fourragere, Versailles.

Johnson, R.A. (1987). *Ecstasy. Understanding the Psychology of Joy*. Harper, San Francisco.

Maturana, H.R. (1988). Reality: the search for objectivity or the quest for a compelling argument. *Irish Journal of Psychology*, **9**, 25–82.

Maturana, H.R. and Varela, F.J. (1987). *The Tree of Knowledge – the Biological Roots of Human Understanding*. Shambala Press, Boston.

Morgan, G. (1993). *Imaginization. The Art of Creative Management*. Sage, Newbury Park.

Rorty, R. (1989). *Contingency, Irony, and Solidarity*. Cambridge University Press, Cambridge.

Russell, D.B. (1992). Social ecology in action, its rationale and scope in education and research. *Studies in Continuing Education*, **13**, 126–38.

Russell, D.B, Ison, R.L., Gamble, D.R. and Williams, R.K. (1989). *A Critical Review of Extension Theory and Practice* (a report for the Australian Wool Corporation). University of Western Sydney (Hawkesbury), Richmond.

Russell, D.B. and Ison, R.L. (1993). The research–development relationship in rangelands: an opportunity for contextual science. Invited Plenary Paper, *Proceedings of the IVth International Rangeland Congress*, Montpellier, 1991. Vol. 3, pp. 1047–54.

Tucker, S.I. (1972). *Enthusiasm: A Study in Semantic Change*. Cambridge University Press, Cambridge.

Watzlawick, P. (1976). *How Real is Real?* New York, Random House.

7 Co-researching: braiding theory and practice for research with people

Lynn Webber

7.1 Introduction

The question of context is central to this chapter. In presenting a more detailed description of what we did in our research, the intention is not to provide a recipe for replication, but to offer insight into how we developed our research framework using theoretically based principles to guide our action. This chapter seeks to paint a picture for those who may consider developing collaborative research processes for similar or other contexts. We also relate how we took our research developed in the context of the NSW Western Division and describe briefly how we adapted it to a novel context. As outlined in Chapter 6, this was part of the expectation built into our research proposal. Our invitation is to read our story and use it as a trigger to develop a research design relevant to the context of your own research.

In this chapter I use the term research to refer to those research actions for which we, the research team, took responsibility. I use the term co-research to refer to those actions for which joint responsibility was taken, although as I will point out, the nature of the responsibility differed between graziers and members of the research team.

7.2 The research region

The region in which this research was conducted is the Western Division of New South Wales, Australia. In these semi-arid rangeland areas we visited grazier (pastoralist) families in transect journeys from Cobar to Enngonia, Wilcannia to Wanaaring and Broken Hill to Tibooburra. Our co-researching community involved graziers in an area around Fowlers Gap, some 110 km north of Broken Hill (Figure 7.1).

7.3 Considering context and valuing diversity

We were in a different world – a context that we could not understand unless we were graziers

As a team of researchers, we placed importance on what we called 'contextual grounding'. To us, this meant *experiencing* some of the physical and social dimensions of what it is to be a grazier in the semi-arid rangelands.

We decided that journeys along three transects would be a useful way to 'ground our work' initially. In-depth interviews were conducted with up to

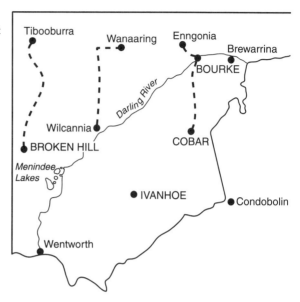

Figure 7.1
The routes of our transect journeys in the NSW Western Division taken at the commencement of our project. (Source: Webber, 1993.)

four grazier families over a three-day period by three pairs of researchers (see Box 7.1).

We began to appreciate the rich diversity of graziers' lives through actively listening to local people's experiences, valuing their perspectives, visiting people on their properties, and driving the long and dusty miles through some magnificent country of outback Australia. Some of the understandings arising from these interviews have been described in Chapter 3.

7.4 Braiding theory and practice

Throughout all our work, both in the development of process and our co-researching with graziers, we made a conscious effort to underpin our practice with theory. Similarly, our practice gave new insights for our understanding of theory.

The theoretical basis which gave rise to our development of principles was that of individual world view (see Box 7.2). Research on perception and cognition supports the argument that each person sees and builds their own reality based on the interpretation of their experiences. This understanding of 'multiple realities' comes from the constructivist position of, amongst others, Kelly (1955), and the biological explanation offered by Maturana and Varela (1987).

The concept of communication on which our process design was based is the 'mutual acceptance of different realities'. We accepted that each per-

> **Box 7.1 Transect visits in the semi arid rangelands**
>
> - Our aim was to **learn** about life in the Western Division first hand from graziers.
> - We hoped to learn by visiting the area, **talking with** people and **listening** to their stories.
> - A **transect** approach, as in Rapid Rural Appraisal (RRA), was used. Three transects: Cobar to Enngonia, Wilcannia to Wanaaring, and Broken Hill to Tibbooburra, were traversed.
> - Local **key informants** were asked to help us identify a diversity of grazier families in terms of property size, family situation, contact with service institutions, number of years on the property, and land system.
> - **Letters and follow up telephone calls** were used to contact and initiate relationships with grazier families.
> - Three teams of two members **visited properties** along the transect.
> - 'Storytelling' (a form of in-depth interviewing) was used to focus on **experiences of life in the rangelands**.
> - If the invitation was made, we inspected parts of the property with graziers.
> - Interviews were **taped** (with permission), **transcribed**, and **themes** developed.
> - Interviews lasted usually 2–3 hours, but were sometimes longer.
> - Two teams of researchers accepted graziers invitations to stay overnight, one did not.
> - After teams departed from properties, **they debriefed to 'let go'** of the last interview. These were taped and included ideas and reflections.

> **Box 7.2 Theoretical principles**
>
> - Invitations to participate are continually extended.
> - Each person's knowledge and experience is unique and valid.
> - Research is not just a scientific approach to 'finding out'.
> - Diversity of approach is valid and useful.
> - Knowledge, experience and understanding gained through action research/learning may result in locally meaningful and adaptive change.
> - Enthusiasm precipitates action, which with critical reflection may result in meaningful action research.
> - Language used is important.
> - Power shared in a collaborative and co-researching sense has potential for mutually satisfying outcomes.

son's view of the world, their experience and how they express it is no more valid than anybody else's.

Theory was useful in that we could develop 'guiding principles', which were integrated in designing and planning 'next steps' in the process. In supporting the research direction taken by graziers, we worked using these theoretically based principles in our action co-researching process.

> **Box 7.3 The research area**
>
> The region chosen for research covered an area of 17 600 km² and included 45 properties, aggregated into 33 holdings. Properties or 'stations' average around 40 000 ha per family unit (although this figure masks the range of sizes). The area also contained a research station owned by the University of NSW (Fowlers Gap), which acted as a logistical base during our visits and was often the site of community meetings. From these, 16 pastoralist families (a sample of 48%) were invited to tell of their day-to-day experiences in interviews conducted on their properties.

7.5 Choosing the area for inviting research action

Following the transect visits we faced an agonising decision as to which geographic area the team would choose as its focus of attention because many valuable experiences had been related enthusiastically by graziers on our visits to their properties. We felt the entire area should be invited to participate in the initial phase of the project but this was not feasible logistically. After much deliberation, the decision was made to invite grazier families north of Broken Hill to join us as co-researchers (Box 7.3).

In order to 'contextually ground' our work in the area north of Broken Hill, and to learn more about the issues and concerns of local grazier families, the team invited graziers to participate in a series of semi-structured interviews. It was not intended that properties selected be based on any criteria other than diversity (physical and social), for example, property size and family background.

As a means of identifying graziers to interview, a grid of sixteen squares was placed on a 75 km radius around Fowlers Gap Station (located approximately half way between Broken Hill and Tibbooburra). From this, properties in each square were randomly selected in order to maintain a broad geographic spread in the sample. Diversity was confirmed using key informants (see page 4 for a map of the research area).

7.6 Designing interviews – bringing theory to life

Here I step out for a moment from a strictly chronological description of our research and focus on the process we followed in conducting interviews. This momentary meander is a discussion of our considerations when designing the interview process with graziers, the most critical way in which our relationship developed with them.

There are basically two types of interview question. A *descriptive* question is one where the response required is a record or description of an action. The second type is where the interviewer invites the interviewee to provide a more sophisticated *analysis upon the raw data of their experiences* – an interpretation of meaning.

The aim of our interviews was not to seek objective 'facts' about events or objects. Our underlying theory pointed to the importance of an individual's *interpretation of their day-to-day experiences* and not on relating facts or descriptions of things which are independent of an individual's interpretation or understanding. We saw as especially important the individual's *pre-understandings*; that is, those understandings from past experiences that shape current experiences. The purpose of interviews was to trigger stories about the interviewee's experiences *and their understanding of those experiences.*

7.6.1 *The role of narrative*

The two main approaches we used to invite graziers to relate their experiences was through semi-structured interviews (SSIs) and a form of in-depth interviewing – storytelling. The common thread in both these approaches was the opportunity they offered to create 'space' for interviewees to narrate their personal experiences as graziers within an interview framework. These approaches provided a focus for the interview but did not dictate the nature of response, rather invited an exploration of meaning. Both approaches were adapted for different circumstances during our research. For example, SSIs were used to invite graziers to relate their experiences of their lives in rangelands. In another situation, they were used in processes of 'critical reflection' on experiences (see below).

Narrative can be thought of as a combination of three aspects: *stories, the telling of them,* and *their meaning.* Stories are of a series of events around a theme as told by the storyteller. In narrative, meaning consists of more than the events alone (as in storytelling) – it also consists of the significance these events have for the narrator in relation to a particular theme (Polkinghorne, 1988, p. 160). For example, this may take the form of the narrator conveying meaning to the listener through a 'moral' to their story, or to substantiate their position on an issue.

Narrative discourse is a dimension of understanding we use in our negotiations with 'reality', especially in terms of time. Many stories follow a pattern of events around themes which are organised meaningfully by a temporal sequence. Plotting of events in a narrative sequence is part of our ordinary existence to the extent that we are usually unaware of it in operation, but only aware of the experience of reality that it creates.

Descriptive narrative research (as opposed to explanatory forms) aims to produce an accurate description of the interpretive accounts individuals or groups use to make sequences of events in their lives or organisation meaningful. Narratives are a recurrent and prominent feature of accounts offered in all types of interviews. If respondents are allowed to continue in their own ways until they indicate they have completed their answers, they are likely to relate stories. Descriptive narrative investigations most often examine stories retained in the fluid oral form rather than the stable written form.

7.6.2 Storytelling as an opportunity for narrative in interviewing

In-depth interviewing is used to parallel the social interaction of a normal conversation (Minichello et al., 1990, p. 117). The nature of the questions asked during an interview influences whether the interviewee responds with a story, as well as giving their analysis. The role of the interviewer in the storytelling process is to create a climate which invites storytelling, allowing stories to unfold with the narrator directing the flow. An interviewer's behavior can encourage the storytelling process in cueing the interviewee that they are receptive to listening to their stories: showing a lack of hurry, engaging in preliminary talk, and inviting detailed accounts at the outset. The narrator is influenced by the environment in an unpredictable way so that what is said is not totally preconceived.

Questioning in an interview needs to be less frequent than would otherwise be the case in ordinary conversation. The entire basis of focus is 'what was the experience and how do you understand or make sense of it?' When necessary, the interviewer needs to bring the narrator back to their experience and not allow flights off into projection about what others have done. It is also part of the process of triggering a positive experience for the interviewee that the interviewer 'actively listens' to their experience.

7.6.3 Semi-structured interviews as thematic narratives

SSIs seek to invite the interviewee to develop a 'rich picture' of their experiences and attributed meaning around themes of interest to interviewer and interviewee. Prior to interviews being conducted, a series of broad themes are developed, rather than a prescriptive set of questions asked in a set sequence, requiring a limited response and to be replicated at each interview. The dynamics of each interview will be different as individuals bring their own experiences and world views into the situation. SSIs have the flexibility for themes to be explored in any order and to create space for narrative in the interview context.

SSIs have been used by a number of researchers in many different contexts to investigate social processes and relationships (Dowsett, 1991). They have been used in Rapid Rural Appraisals (RRA) to identify issues of rural people (Beebe, 1985). We saw the thematic nature of SSIs as potentially useful for our process for inviting narrative within a focused interview framework.

7.6.4 Generating our own understandings of these approaches

In designing the interview process, we sought to generate our own understandings of how narrative could be invited through storytelling. A tension throughout the research was the different perspectives of the researchers on whether semi-structured interviews (SSIs) or in-depth interviewing created the richer context in which stories might be triggered. Ray developed a thematic narrative metaphor of a river in describing the dynamics of the SSIs as 'each story flowing and meandering along its course'. He considered the skill of the interviewer to be critical and used the metaphor as a means to monitor the dynamics of the interview and to maintain mental notes of 'tributaries that one sensed had entered the river system which could be returned to when the flow of the present stream had ebbed'. Other researchers felt that SSIs, particularly if not conducted competently, predisposed one to eliciting the 'facts of the matter' rather than stories. There was much discussion around the notion of 'patterns' as a characteristic of quality (measurement being seen as a characteristic of quantity). Two main types of patterns emerged: *on the ground patterns*, such as space, time, flow and decisions (as in Conway, 1983) and *meaning-making patterns*, such as those that derive from stories. The relationship between the two types of patterns constituted a third, connecting form which was complementary in building a 'richer picture'.

Emerging from this reflection on patterns, narrative and interview approaches was an understanding of the role of narrative for our research. In both in-depth and semi-structured interview frameworks, there was potential to invite narrative in less thematic and more thematic forms, offering insight into the patterns of meaning made by graziers. SSIs could also be used to focus on eliciting data to develop on-the-ground patterns. With SSIs, we had a conscious view of what that particular interaction may offer in terms of understanding, and postulated that it could be a useful way of triggering enthusiasm, although in-depth interviewing through inviting storytelling was seen as a more open-ended approach with possibly greater potential to allow us to develop methodology around the concept of enthusiasm (see Chapter 6).

We felt that the experience of having told a story is in itself a 'change-making' experience. Expression of stories allows people to be seen as a resource for decision-making and ongoing insights, not only as individuals but possibly for the broader community. The sharing of stories allows for the acceptance of multiple realities. The mutual acceptance of these differently experienced realities also offered a potential basis for community development.

> **Box 7.4 Interview protocol**
>
> 1. **Setting the context**
>   Go **with the flow**, with respect to who the interviewee focuses on, prefers to be interviewed by (if in pairs), where to sit, etc.
>   Prior to the interview there may be an opportunity to '**break the ice**' over a cup of tea, talk generally and explain the introductory part of the interview.
>
> 2. **Introducing Ourselves**
>   **Explain who you are**.
>   Use a **calling card** if appropriate.
>   **Refer** to the **phone call** made to confirm the interview.
>
> 3. **The Letter**
>   In reference to the letter sent, elaborate on this **by inviting any questions** about what you are doing.
>   Some of the questions asked may relate to research funding, who you work for, how they were chosen for interviewing, why you are interested in talking with them. Be clear and prepared to answer these questions.
>
> 4. **Where We Stand**
>   This relates to the **theoretical position** on which the research is based:
>   - Our project is concerned with developing more effective R&D which better meets your needs.
>   - You have knowledge and experience that we value and you can share with others.
>   - We are outsiders and you know more about your region than we do.
>   - We invite your help with our research.
>   - We cannot promise anything but we are prepared to work with you.
>
> 5. **Why we are here talking**
>   - We would like to talk with you about your **experiences** as a grazier (pastoralist).
>   - With your permission we would like to **record** the interview which will remain **confidential**.
>   - Indicate who will **conduct** the interview and who will record.
>   - Restate the minimum **time** commitment for interview.
>   - How do you feel about that?
>   - Anything you would like to **ask** us before we start?

7.7 Conducting the interviews

7.7.1 *An interview protocol*

We gave much consideration to the 'relationship building' process preceding actual property visits for interviews (Box 7.4). Grazier families were contacted firstly through personal letter, followed by telephone conversations to discuss any aspect of our invitation prior to visiting graziers on their properties. We developed a protocol prior to conducting interviews which attended to the process of introductions, inviting question of clarification, and conducting interviews.

7.7.2 *Semi-structured interviews for triggering enthusiasms*

SSIs were used in our initial visits to grazier families in the selected area. These interviews focused on graziers' stories of day-to-day experience, anchored in being a pastoralist, across three contexts in time: the past, the present and the anticipated future.

Inviting exploration of an individual's tradition (or past) recognised that our pre-understandings shape our current experiences. Anticipating what daily life will be like seeks to explore the future context. One of our main desired outcomes from our interviewing of graziers was for them to experience having been actively listened to, by having what they said accepted as worthwhile, and to be left feeling valued as a person.

Our enthusiasm to hear about pastoralists' day-to-day experiences was supported through techniques such as eye contact (seating position, showing interest and picking up non-verbal signals), body language and positive feedback (verbal and non-verbal). Questioning was minimal and broad to invite stories. The team member accompanying the interviewer was an active listening participant in the process.

In rounding off the interview, we moved towards more general questioning to dissipate the intensity of the discussion. Acknowledgment of the conversation was an important part of the valuing process, as was our specific thanks to them for spending time with us. Before leaving, we invited the interviewee to ask further questions of us, to ensure that graziers were left in a position where they could set the focus of discussion. This often led to further fruitful conversations.

7.7.3 Semi-structured interviews for evaluating and relating experiences

The nature of SSIs in evaluation took the form of both individual and group interviews, conducted in private with a team member, or in public, as a way of graziers relating their learning experiences to other graziers.

SSIs were conducted in the lead up to, and following specific action events associated with our co-research. We invited graziers to discuss their perspectives prior to taking action – how they saw the situation and why. Graziers also talked about the sorts of questions they were asking. Following an event, graziers were invited to critically reflect on their new experiences and the insights or understanding gained from them. SSIs were also used in a process of public interviewing of graziers as a means to generate public data for evaluative purposes. Researchers interviewed graziers with other graziers listening as an audience. These data from the interviews were combined in a matrix (see below) and following collective discussion a final form for the evaluation matrix was agreed.

7.8 Making sense of semi-structured interview data

The data from the SSIs were used in two ways: (i) the development of thematic posters for use in a 'mirroring' process in a community workshop which followed the first round of interviews (Figures 7.2(a–(d)); and (ii) full transcriptions of tapes, which were compiled for later analysis.

170/A design for second-order R&D

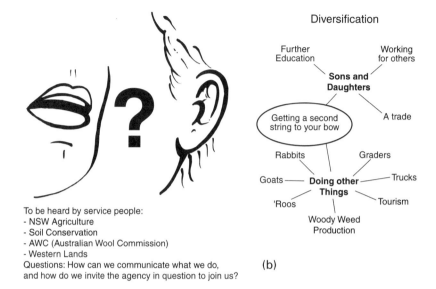

To be heard by service people:
- NSW Agriculture
- Soil Conservation
- AWC (Australian Wool Commission)
- Western Lands
Questions: How can we communicate what we do, and how do we invite the agency in question to join us?

(a)

(b)

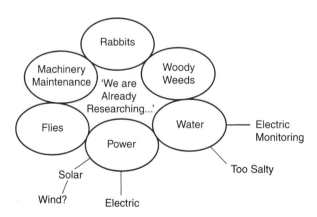

(c)

Figure 7.2(a)–(d)
These figures depict thematic posters developed using SSI data. These were used in the community workshops. The posters prepared represented our interpretation of what we heard, but were in no way intended to be a synopsis of local community perspectives. The statements used in the posters were data sourced from the SSIs and quoted anonymously. Our use of grazier's data in their language in the process of 'mirroring' sought to act as a 'trigger', with the potential to create new insights or ways of thinking about these issues.

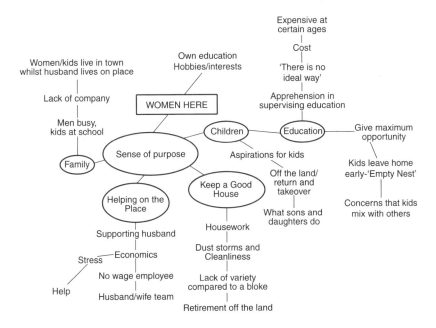

(d)

Maturana (1988) recognised one way of triggering change in human social systems as being through interactions which invite people to reflect upon situations as they are currently understood. The idea of a 'mirror' is that it does not reflect a 'reality' but an image, or interpretation of it. What we sought to do was to offer our interpretations of graziers' situations based on the understandings and enthusiasms which each of us as 'researchers' derived from participating in the interviews. These were used to: (i) demonstrate that we had listened to, and heard the concerns of, graziers; (ii) act as 'trigger' material for graziers when invited to reflect on their own situation and to name research issues of concern to them (see also Webber and Ison, 1994).

The data from the SSIs were used in the development of thematic posters for the community workshop to which graziers in the focus area were invited following the first round of interviews. Meaning was attributed by individuals in the team to the data obtained through our SSIs with graziers. We were therefore engaged in a process of 'constructing' what we interpreted as graziers' perspectives. The choice of themes was left to individual team members, based on the issues they were most enthusiastic to reflect to workshop participants (Figures 7.2 (a)–(d)).

Themes were developed by linking similar and related issues (or sub-themes) together and building an overarching theme which would provide an interpretation of what the team had learned from actively listening to grazier families. The four posters which are depicted reflect: the widespread concern held by graziers that their voice was not heard (Figure 7.2(a)); that to survive in this physically and socially demanding environment graziers need a diversity of activities and income sources – 'a second string to your bow' (Figure 7.2(b)); that in terms of 'traditional research' graziers were already active experimenters (Figure 7.2(c)); and that women who lived in this environment did so often under intense pressure and hardship and faced many demands which shaped their sense of purpose (Figure 7.2(d)).

7.9 The workshop design

We extended an invitation to local grazier families to participate in a workshop after conducting the semi-structured interviews (SSIs). This was in a community setting and sought to explore enthusiasms for action around local concerns or issues (see Chapter 6).

The workshop provided opportunities for local people to meet the whole team in person, to meet with other local people in the context of the project, and to question and clarify the project's aims. Participants were invited to listen to our interpretation of the main concerns and issues we had heard during the SSIs conducted over the preceding four days. Local people were offered 'space' to explore their issues, concerns and enthusiasms for action. For the team, the meeting was an opportunity to meet with a larger number of local people than those interviewed and, in conception, provided a potentially broader base for community action.

Our experience on this occasion was, however, that the attendance was largely restricted to those graziers who had participated in the interviews – where some form of personal relationship had already been made. This, of course, raises questions about what constitutes, and is experienced as, 'an invitation'. Our experience here, and in subsequent research (Kersten, 1995) suggests that relationship, building is a pre-requisite to on-going collaboration when no single issue has galvanised a community.

7.9.1 *An overview of the workshop process*

The team devoted a lot of time and energy to developing a process which reflected our theoretical position – providing a trigger for local people's enthusiasms for action in a way which was sensitive to the context (that is, available rooms, number of participants, time available, etc.). After many

> **Box 7.5 The workshop stages**
>
> 1 Introduction
> 2 History of principal researchers
> 3 The project approach to research
> 4 Mirroring
> 5 Break
> 6 Small groups session: comments, questions & clarifications
> 7 Small groups session: ideas and interests around which action might be taken
> 8 Break
> 9 Plenary session: review of your suggestions
> 10 Clarification & confirmation
> 11 Vote with your feet
> 12 Statement of intentions

hours of discussion and planning, a twelve-stage workshop design was prepared, which formed the basis for an agenda which we invited the participants to consider (Box 7.5).

The workshop design included a series of stages because we considered the process needed to explore the project history (that is, the story of the researcher's background), the theoretical basis of the project and how this contrasted with traditional approaches to research, and to incorporate the 'mirroring' of our understandings. We sought to provide a sufficiently rich experience for the graziers, who had little (if any) exposure to researchers and who perceived research and development in very conventional terms.

The workshop was conducted at Fowlers Gap Research Station, located 110 km north of Broken Hill. The team invited (by letter and follow-up phone conversations) all forty grazier families from within the 75 km radius around Fowlers Gap. Some people attended from properties more than one hour's driving time away.

7.9.2 Introduction to the workshop

A brief outline of the workshop process was given in the form of a proposed agenda (Box 7.6). The outline also contained some broad timings so as to indicate the commitment required of participants during the workshop, breaks and the time that the social activity would commence.

Participants were invited to change the agenda to meet their needs. At the outset of the workshop participants were invited to interrupt and ask any questions for clarification as the workshop proceeded. We also sought and gained permission to take photographs, or to use tape recorders.

At the start we suggested that the meeting would not be conventional for the participants, who were more acquainted with the traditional meeting procedures. That this was a first experience for many was recognised. The finalised programme was left on the wall for ongoing reference as the

> **Box 7.6 Program**
>
> - Introduction
> - Background to this research
> - What we have learned from you
> - Questions & clarifications
> - Ideas & interests around which action may be taken
> - Break for drinks
> - Review of your suggestions
> - Summing up

workshop progressed. There was no assumption that people knew the process and as much as possible the language used by the facilitators was kept to the language used by the participants (an added advantage in conducting SSIs during the week prior to the workshop). The extent to which this was achieved, despite our efforts, is revealed in the comments of the graziers, in Chapter 8.

7.9.3 *Relating our histories*

The principal researchers, David and Ray, told the story of their explorations and experiences which had brought them to the project. This was seen as a means of relating the project to the context of where the researchers 'were coming from'. Our aim was to provide an explanation as to why we had enthusiasm to get a project like this going and why we were now talking to them, so far away from our base in Sydney.

7.9.4 *What we see as 'research'*

We endeavoured to clarify the project's stand on the meaning of 'research', in a way that was meaningful to participants. This was undertaken by personalising the definition to graziers' situations:

> 'Research' is about investigating something which makes a difference to me (the grazier) and which can be incorporated into community knowledge for our benefit.

Our notion was that three types of research could be recognised. We suggested that this project was different because it fitted the third theoretical framework, that of researching *with* people.

> Research on things (plants and animals)
> Research on people
> Research with people

We described the project as a community approach to research, whereby researchers like us, from institutions, work *with* people on issues they see as important rather than on issues researchers, as outsiders, see as important. In terms of research, graziers were invited to be the 'driving force' with (or even without) other researchers, dealing with funding bodies directly, indirectly, or taking other steps.

We contrasted our approach with the traditional 'transfer of technology' research process which was familiar to them:

Funding Body → Researcher → Extension → Grazier

We explained that we were not working in the traditional sense of agricultural research and development. Rather, as encompassed in our definition of research, we were taking a much broader perspective than they had been used to. We sought to appreciate that 'research' reaches greater dimensions: there is a whole life out there, of which woolgrowing is one part. We suggested that there was nothing wrong with people having enthusiasm to pursue research issues other than sheep, dams and fences if those issues were considered of overriding importance at that time. For example, we had clearly heard concerns expressed in the SSIs over educational opportunities for children and how this influenced the management of financial resources on some properties. We presented a theme in our process of 'mirroring' on production issues (rabbits, woody shrubs, etc – Figure 7.4). This triggered many of those graziers present to say they could look after those issues, and that other issues such as wool marketing was what they wanted to pursue!

7.9.5 *'Mirroring'*

The approach we developed acknowledged that people should not be divorced from the process of assimilating and making sense of data in their own right if they are to come up with themes or issues which they could claim as their own. The 'triggerer' mirrored (provided) personal perceptions. The presentation was an outsider's interpretation which belonged to the outsider because it had been processed and articulated by them.

We considered the process of mirroring to be an important stage which had the potential to provide creative insights and which might enable a reconceptualisation of individual concerns. The importance of the presentations was to 'trigger' new ways of seeing the issues presented (or other issues). This reconceptualisation of issues was seen as a precursor to taking effective action.

All the team members participated in this part of the process. This involved a three or four minute presentation on each poster. The languages

> **Box 7.7**
>
> Following the presentations by team members, time was allocated to a break, giving participants an opportunity to gather informally and discuss the content of the presentations or to study the posters more carefully. This stage was an important component of the process as it gave time for the participants to consider the perceptions presented by the team. A grazier who had been involved with local community action in the past commented that we were in touch with issues that had taken him years to appreciate.
>
>
> *What does all this mean for me?*

of the participants was used as much as possible, using where possible direct quotes from the SSIs to illustrate an issue. Following presentations, the posters were pinned up around the room to enable people to examine and discuss them over the break which followed.

7.9.6 *Having a 'break'*

The 'mirroring' process certainly 'worked' in terms of demonstrating that we had listened to, and heard, what graziers had had to say. There was a great buzz of discussion during the break. It also worked in terms of providing the space for graziers to consider research in different terms, and which subsequently allowed them to name higher order issues than those which had normally been associated with research and development in this region. For almost all those present our distinction about research articulated something they sensed intuitively – in informal discussions during the day and evening these distinctions were often returned to by those present.

7.9.7 *Small group sessions*

The small group session comprised two distinct parts: 'Time For Comment and Clarification', and 'Ideas and Interests Around Which Action Might Be Taken'. Family members were invited to join different groups supported by a facilitator to guide the proceedings. These sessions were taped and photographed with the permission of the participants.

The first stage provided an opportunity for participants to question, comment on and clarify what had been presented in the team's mirroring activity. The structured format of the session provided the opportunity for equal contribution from all participants. It was conducted within a specific time frame to ensure people were aware of how long they were allocated. The facilitator initially emphasised that participants were part of a group and everyone had the opportunity to have a voice.

> **Box 7.8 One issue nominated by participants**
>
> **Issue: Education**
> Video library on specific problems, e.g. constructing a windmill.
> Tertiary education for children and adults.
> Why has the tax deduction for the education of isolated children been removed?
> Further education for women & children, e.g. tele-conferences.

The second part of the small group session involved reissuing an invitation to those present to nominate their 'research' concerns. The process was less structured and involved a discussion of ideas and interests around which action might be taken, encouraging contribution from everyone.

At the conclusion of the discussion, participants were invited to record each of the issues that had been discussed in the small group. This was written on poster paper large enough to be seen from a distance when pinned to the wall of the meeting area. Participants were invited to write their issues or dictate statements for the facilitator. An example of one issue was around the theme of education (Box 7.8). By this stage the team members' interpretations of local peoples' concerns and issues was considered redundant. The graziers present had generated issues which belonged to them and expressed them in their own language. This stage in the process resulted in a collection of statements on individual pieces of poster paper. These were then pasted randomly around the room for the other groups to view and discuss.

A break in the workshop programme followed this session in order to allow participants to view other groups' statements of issues. During this time participants moved around the room viewing others' contributions and discussing the outcomes. Participants were encouraged to write additional issues which had not previously been thought of on the blank poster paper available.

7.9.8 *First plenary session*

The small groups came together for a larger 'plenary' session in which team members introduced the issues generated in the small groups, referring to the issues pasted around the room. Each issue was further developed by participants who were invited to annotate or add to any of the posters that described an issue they felt strongly about. Thus, all the participants had the opportunity to learn what others thought were issues which they might consider for 'research action'. Whilst this was occurring other team members were grouping, or 'nesting', issues on a master sheet in preparation for the next step in the process.

The workshop adjourned to a BBQ dinner with the team as hosts. This gave the participants the opportunity to talk amongst themselves before the second plenary session. During the BBQ, team members reorganised the issues generated from the small group discussions. The individual issues were 'nested' into similar issues. For example, issues such as tertiary education, school and further education were nested as 'education'. Other main nests included wool marketing and interest rates. People were encouraged to look at these whilst the BBQ was in progress and talk with each other and the team members. Clarity about the appropriateness of the 'nesting' was sought from the participants. (This stage might well have been conducted by the participants themselves – but that was our learning from the experience.)

7.9.9 *Second plenary session*

Participants were formally invited to confirm the appropriateness of the 'nesting' of their issues. Again, this sought to reaffirm ownership by those who had generated the issues. We invited agreement or disagreement with the arrangement of the issues in their nests, which resulted in moving issues around until participants were happy with the groupings. If an issue fitted into more than one nest then there was scope for that to happen. This recognised the inter-relationships in complex systems where one item does not necessarily fit comfortably into just one box.

There were five major groups formed around different issue 'nests' – education, retirement, interest rates, government interference, and collecting local stories. Workshop participants were invited to gather around any 'nest' they felt enthusiastic to take action on. The invitation was to 'vote with your feet'. An equally valid choice was not to have sufficient enthusiasm for action on any of the issues generated in the course of the workshop (Box 7.9). Participants joined others for discussion around their particular issue of interest.

The outcome of this process was to establish whether there would be a 'next' step (whatever that would be). There was no expectation that people knew what they were going to do next, more the point that they were going to *DO SOMETHING* next (i.e. take action).

Those in the group were invited to give an indication as to whether they wished to push on, and whether they wished the research team to return to support them in their research. The reply was a unanimous yes: the graziers wanted the team to come back and help them with the process. The names of the interested people were grouped and recorded under their issues of

> **Box 7.9 The notion of enthusiasm**
>
> Enthusiasm has several dimensions: theoretical, emotional and methodological. Theoretically, enthusiasm is derived from Greek words that mean 'our understanding comes from within ourselves'. Enthusiasm as a methodology can be seen as shaping our approaches to interaction in such a way that we do not divert a person's energy, but explore where a person's energy is. This requires respect for the person themselves as they are. Inviting people to express their personal enthusiasms for action was the basis for the workshop process. We see enthusiasm as already present and not something that can be instilled by others.
> Enthusiasm will cover a broad spectrum of potential activities.
> When people with similar enthusiasms for action come together, there is potential for satisfying work together, if there is enough energy and willingness for action. This cannot be determined by any other person.

interest. One participant in the workshop did not identify with any issue and legitimately opted to remain uncommitted.

7.9.10 *Concluding the workshop*

Team members thanked all the participants and caterers. We said that we believed there was sufficient interest in what it was we were doing to come back. There was some tension with some graziers who were concerned to know what our 'real agenda' was. Our approach to research with people was certainly alien to their past experiences of researchers and their 'messages'.

The team members made an undertaking to correspond with people. We agreed to return at a time when it was best for graziers to come together. In the meantime, a report of action to date in the form of a newsletter would be sent to local grazier families in the area in order to share the outcomes thus far and invite further participation at any stage. This ended the proceedings for the team and made a clear close to the workshop.

7.10 **Extending another invitation**

Following the first workshop, it was suggested by the participants that the team return for another workshop to invite more local grazier families to be involved. Participants in the first workshop had volunteered to talk with other people in the area about this workshop, and invited them to attend. Local people took responsibility for helping the team organise the workshop. We sent a newsletter to all local people with a report on the first workshop and the news that the team had been invited to return for further work by the local people who attended that workshop.

The second workshop process was similar to the first in its initial stages. Based on reflections on the first workshop, we felt that in this workshop we needed to support graziers' triggered enthusiasms for action towards developing a more concrete focus for action. Having a more tangible focus could also assist in explaining to other people what the project was about.

In the small group session, team members facilitated a 'scanning session' using a 'bus driving' metaphor to introduce the session: *I see that research is like driving a bus – at the moment the driver is the scientist. The future could be graziers as drivers or co-drivers of research with scientists.* The intention was to invite participants to think about potential co-researching relationships. Participants were invited to consider the following questions: *What is happening in the world around you? What do you think is likely to happen? What would you like to see happen?*

Participants then discussed and recorded their responses to these questions. Some people preferred to work on their own, whilst others discussed the questions among other members of the group and recorded their responses themselves or as a group.

The next step involved development of action statements. Team members invited participants to think of 'doing' statements for issues and what their outcomes might be. Ideas for action were not discounted but the focus was on '*what*' rather than '*how to*'. These statements were developed as:

I want to do (find out about) something that will be of benefit to me and possibly others in the community.

At the conclusion of the second workshop, graziers identified the priorities for action they were most enthusiastic about. Graziers focused on two major issues: wool marketing and community building.

When we commenced the workshops we had no preconceived ideas about what research would be done – we certainly were anxious, but tried to hold these back in our interactions. The decision to research wool marketing could be seen as an ideal outcome in terms of our funding source – but it was the outcome of the process and could have legitimately been other issues, such as health or education.

On reflection, it is at this stage that our overall design lacked cohesion. Our process had clearly triggered enthusiasms, although it was not clear to us what processes had occurred between workshops one and two which resulted in a reduced number of enthusiasms being articulated. In conceptual terms this period reflects the switch from enthusiasm to concensus (see Chapters 6 and 9). What was clear to us was that as co-researchers we would undertake to support graziers in their research action with the graziers driving the direction of the research action and the team offering their support as facilitators of action. However, logistically and in this context (both physically and institutionally) it was impossible to facilitate co-research on all of the enthusiams that were generated.

The concept of inviting graziers to join us as co-researchers in a process of action research was not one that was easy to imagine, and was certainly an invitation to 'take the plunge' and work in a new way with us. The graziers needed time to think about our discussions and what opportunities this provided. *Continuing the conversation* was an important aspect of developing understanding about what we were trying to achieve. This involved many iterations. We thought it important that as researchers we had some form of vision of what might be possible, whilst holding back from prescribing action and thus taking away the opportunity for graziers to be responsible (by naming specific actions to be taken) for this co-research. In practical terms, 'continuing the conversation' meant maintaining contact and talking, clarifying and reassuring. This was difficult logistically, and relied on a lot of phone conversations, mainly at night after graziers had finished their day's work.

7.11 Action co-researching

A process of action research around the issues of *wool marketing* was initiated. In our terms, following Maturana and Varela's (1987) notion, the 'community' included those people engaged in the 'network of conversations' – grazier families (men, women and children) and the team members. Within this framework, the concept of 'outsiders' (the team) and 'insiders' (graziers) does not apply – both undertake a learning process together.

It is not the purpose of this chapter to describe in detail the co-research subsequently conducted (see Chapters 6 and 8 and CARR, 1993). The specific actions are detailed in Table 6.1. Our co-research can be seen as an elaboration of the action research or **experiential learning** cycle which follows an ongoing process of **planning, action, observation** and critical **reflection** (Box 7.10).

In terms of group process, subsequent 'research' meetings with the grazier community provided a framework in which to take action. The location for the meetings was decided by the graziers, and more often than not, were on different properties (stations) in turn. This provided a familiar context for those involved. It was also equitable in terms of the distance that had to be travelled by individual graziers. On-going facilitation by one of our team was critical at this stage.

7.11.1 *Graziers relating their research experiences*

An important dimension to the action co-researching process was the opportunity for graziers to relate their individual experiential learning, the **local knowledge and experience of graziers developed over generations**. The co-researching process sought to value these experiences, and new

> **Box 7.10 The action research cycle**
>
> Action research had its beginnings with Kurt Lewin, who saw it as concerned with formulating and solving problems through a series of 'cycles' of plan, act, observe, and reflect. The aims are for individuals to learn by doing, and through experience gain insight and understanding. This may lead to improvement in practice and alter existing situations and constraints.

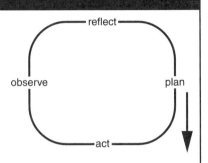

> **Box 7.11 A story of taking action**
>
> This story is about graziers visiting a woollen mill to gain 'first hand' information about direct selling their woolclip to the mill. This is an example of graziers 'finding out' about issues around wool marketing that they were keen to understand more about.
>
> **Some comments before the visit...**
> 'We believe today that we can try their system and really have nothing to lose. We have an idea about how they price wool and the main consideration would be to be able to accept a price without feeling at a later date that you had been ripped off.'
>
> **Some comments after the visit...**
> 'I was really pleased to hear that the producer can sell direct to a mill and not be taken advantage of. It seems this doesn't happen when you sell to a private wool buyer.'

experiences and the learning to come. By individuals **relating** these experiences through critical reflection, community knowledge is built and is embedded in subsequent planning, action, observation and reflection.

At the 'research' meetings all those involved were invited to tell their stories of 'what happened' when they took action on something. This was a basis from which others could ask questions to create a learning network. The *critical* dimension of reflection was focused on 'meaning making' from experiences. The next step evolved out of this understanding – be it to follow some new line of investigation, or a different angle or to drop it altogether! Again the role of the facilitator was critical as a facilitator of learning and particularly, critical reflection – what did it mean? The other great skill required by the facilitator was to avoid taking responsibility for the research action (which it is possible to do in many ways, such as suggesting lines of inquiry, contacts, meanings based on the facilitator's experience, etc).

7.11.2 *Experiences and learning*

All along the way we were learning from our experiences in the explorations around the issue of wool marketing. Even the 'dead ends' and the 'not

Table 7.1
Systemic action research

Learning	Experiences
• Local pastoralists gain first-hand knowledge • Take responsibility for action and outcomes • Sharing power • Mutual benefit • Better at researching their needs	• Interviews • Workshops • Visits • Meetings • Ongoing communications with service institutions

useful' experiences helped create a richer picture of the issues at hand. Relating critical reflections on our experiences along the way took place not only within our co-researching community, but with the wider local community. Collaboratively developed newsletters following key events, radio interviews, and excerpts of information gathered by local people were all used to relate stories of our action with others.

The primary purpose of the newsletters was to convey the sense of invitation – that others could join if they wanted to, and that they were not being excluded in any way. This appeared to have been the case, based on a round of exploratory conversations we conducted in the Fowlers Gap area mid-way through our research, but did not lead to greater participation. People valued being in touch, and were happy to see those graziers involved taking action on what they generally saw as an important issue. The relationship between learning and experiences in this systemic action research are summarised in Table 7.1.

It was important (to us as researchers) to document the learning from our co-researching experiences around the wool marketing issue. In doing this we sought to provide a community resource and a resource for service institutions or other research communities. In terms of community building, relating experiences both within and outside our co-researching group formed an integral part of the ongoing relationship building process.

The process of documenting experiences was via collaboration with graziers who had undertaken particular action. These articles took the form of graziers relating their stories of their experiences through reflection SSIs conducted on both a one-to-one basis and in group meetings. Some documents were developed through several iterations of drafts which were 'given back' to graziers for comment and clarification (e.g. Chapter 8).

7.11.3 *Invitations to reflect and relate experiences – evaluation*

During our co-researching process, invitations to reflect on experiences and relate perspectives offered an opportunity for all of us to consider our understanding in the light of other peoples' perspectives. The use of public

SSIs provided a context for this to take place. These SSIs were initially facilitated by team members and later by graziers themselves, interviewing other graziers whilst others watched and listened.

In the course of taking action, perspectives and questions were recorded before and after specific events, for comparison. This sought to provide a resource for other graziers, and to identify experiences enhancing understanding throughout the process. It also provided an evaluation of our co-research action which was publically accountable and not confined to meeting the evaluation demands of any one stakeholder in the R&D process.

Conventional evaluation criteria require pre-determined measures for pre-determined outcomes. This is incompatible with co-researching design and participatory processes which, because it is second-order R&D, requires second-order data as a basis for evaluation. The 'outside' nature of first-order criteria encourages measures of success to be determined externally to the process itself. This takes evaluation away from the context in which the co-researching process is taking place.

We maintained ongoing 'conversations' with graziers who were taking action, actively listening and inviting graziers to relate their perspectives on new insights or experiences in a critically reflective way. This dimension of evaluation sought to focus on *what 'research' was for graziers: doing (finding out about) something that is of benefit to me and possibly my community.*

Modes of evaluation were designed as part of the process itself. These included the conduct of personal SSIs in the lead up to, and following specific action events. We sought to focus on grazier's critical reflections in evaluating their own experiences and the insight or understanding gained through them. Public SSIs involved individual graziers being interviewed with other graziers as audience. This process sought to offer graziers an opportunity to experience SSIs from another perspective to that of the interviewee, and to create a context within which graziers could relate their understanding and insights gained from experiences to others through critical reflection. An evaluation matrix was developed which showed process-based criteria of 'driving force/concerns', 'action' and 'usefulness' across time (past, present and future) (Figure 7.3). This template has been found useful in a number of subsequent contexts in which it has helped bring into the public domain data which may often be inaccessable because of predominantly oral, as opposed to written, traditions.

Table 7.2 provides some data relating to participation in our research (defined as active involvement in the process) in terms of 'first-order' data. These data are derived from our research in the Fowlers Gap area and in the Kyeamba Valley in the south-western wheat–sheep zone of NSW where our workshop design was also tested (see below).

Table 7.2
Participation statistics in two regions in which our initial research design was tested

Number of graziers/families interviewed		Number of workshop participants		Number of participants not interviewed		Number of landholder families in the study area	
FG	KV	FG	KV	FG	KV	FG	KV
16	30	14	72	2	42	33	110

FG, Fowlers Gap; KV, Kyeamba Valley.

	Past	Present	Future
Driving force or concern			
Action			
Usefulness			

Figure 7.3
A template for use in public semi-structured interviews for the purposes of evaluation, through critical reflection, of research action. (From CARR, 1993.)

The data suggest a robust process although the initial participation in any given population may not be greater than in traditional R&D approaches. However, our research was not designed to manipulate people into participating nor was it designed to emulate the type of relationship which has historically existed, and often continues to exist, between service institutions and local people. It was designed to trigger enthusiasms for R&D action. In the Kyamba PRA, initial participation in the community workshop was greater than twice the number of families interviewed, in contrast to the research reported above (e.g. Webber and Ison, 1994). What differed was the institutional and resource (including logistical) context for the research and subsequent evidence for on-going co-research in the case of the Kyeamba PRA.

7.12 Community capacity building

From our mutual experiences as co-researchers, we are able to say that graziers have worked in a different way in researching their issues and concerns. We feel we will have been successful in developing effective learning communities when local people no longer need us to support their enthusiasms for research, but have sufficient capacity to 'do it' themselves. Our research did not continue in ways that enable us to affirm or deny this (see Chapter 9).

Our redundancy as research facilitators would mean that local people have the capacity for ongoing individual and collective research action – that the knowledge of a way to go about this is held by the community and can be offered to others. Having participated in our project means that a way of working and learning has been experienced by those involved and the potential exists for the approach to be used again if needed.

Empowerment of local people, if it has any meaning at all as a concept, must relate to the control of processes: ownership of design, action and outcomes from which they will benefit and which have the potential to be used in novel contexts. The learning from action research of our co-researching community was 'first hand'. Those involved heard directly from the people who could answer the questions asked – their own meaning or interpretations could be made. The opportunity to enter into conversation with those who could answer the questions meant that answers which were not sufficient for people's needs could be clarified or expanded upon or reshaped on the spot. The learning culture of research communities is not necessarily solely through written material. Our researching community found it much more useful and rewarding to enter into conversations with people. This all requires quality time with people to allow for meaningful insight.

7.13 Continuing our research journey

Our research journey has taken some meanders along the way, which have all contributed to our understanding of theory, process design and practice. These experiences have included the writing of papers for academic audiences in conferences and journals and the design and conduct of workshops with research and extension managers and practitioners. The initial process of inviting landholders to express enthusiasms for action was adapted in another context as stipulated in our initial research proposal (Chapter 6).

7.13.1 *Adaptation of the process in another context*

Two team members were invited to collaborate in a Participatory Rural Appraisal (PRA) in the Kyeamba Valley, a mixed farming area of southern New South Wales. We were keen to adapt the workshop process designed by the CARR team using the same theoretically based design principles to those used for designing the Fowlers Gap workshops (Webber and Ison, 1994).

Following a series of semi-structured interviews with some graziers an invitation was extended to all local landholder families to come together to explore enthusiasms for action. The Kyeamba Valley people were geographically closer to each other and the workshop process involved a greater number of participants. However, in convening small-group sessions in the process, we recognised the importance of listening to and valuing every individual's perspective in a context which invited this.

The 'nesting' of issues step was adapted to deal with the large numbers of issues which participants had offered, broad issue groupings were supported by a summary of the key features of that issue as articulated thus far. A team facilitator was introduced for each issue group and participants were invited to join in discussion in small groups with people who had similar or related concerns. There was no obligation to participate or to join the group which related to a participant's nomination on paper. A second small group session followed, which focused on identifying and exploring opportunities for action. A plenary session then offered a forum for all issue groups to offer their ideas and intentions to all workshop participants.

Inviting commitment to ongoing action was a feature of the final session of the workshop. This included local people undertaking to think about what had happened at the workshop, talking with neighbours and exchanging phone numbers with others in their issue group. Others generated specific action statements, such as meeting again to discuss direct marketing. The nature of action statements indicated that local people required more time for discussion and planning. Within the time constraints, the workshop based on the CARR design to a large extent served to set the scene for ongoing action in the future. On reflection, we found our design robust, but in keeping with our theory it was adapted to meet a new context and not followed like a recipe.

Acknowledgments

Many people offered their ideas in the development of this chapter: the graziers in the Fowlers Gap area – our co-researchers; the project team: Ray Ison, David Russell, Peter Davey, Philippa Major, David McClintock and Danielle Dignam. I especially thank Philippa for our discussions on narrative and storytelling; and Stephany Kersten, The University of Sydney, for her constructive suggestions in the writing up phase of this work.

References

Beebe, J. (1985). Rapid rural appraisal: the critical first step in a farming systems approach to research. *Farming Systems Support Project Networking Paper No. 5.*
CARR (1993). *Marketing of Middle Micron Wool: Researching With People on Issues that Make a Difference.* Monograph, Community Approaches to Rangelands Research (CARR) Project. University of Sydney and University of Western Sydney (Hawkesbury), Australia.

Conway, E.R. (1983). *Agroecosystem Analysis*. ICCET, Series E, No.1, Imperial College, London.

Dowsett, Gary (1991). Interaction in the semi-structured interview. Paper presented at the CARR Workshop. Macquarie University, Australia.

Kelly, G.A. (1955). *The Psychology of Personal Construct*. New York, Norton and Co. Inc.

Kersten, S. (1995). In search of dialogue: vegetation management in Western New South Wales. PhD Thesis, University of Sydney.

Maturana, H.R. (1988). Reality: the search for objectivity or the quest for a compelling argument. *Irish Journal of Psychology*, **9**, 25–82.

Maturana, H.R. and Varela, F.G. (1987). *The Tree of Knowledge: The Biological Roots of Human Understanding*. Shambala Press, Boston.

Minichello, V., Aroni, R., Timewell, E. and Alexander, L. (1990*). In-Depth Interviewing: Researching People*. Longman Cheshire, Melbourne.

Polkinghorne, D.E. (1988). *Narrative Knowing and the Human Sciences*. Albany State University of New York, New York.

Webber , L. (1993) *Design and Conduct of Co-researching Processes . . . Braiding Theory and Practice for Research With People* (Monograph). 24 pp. CARR Project, WRDC, Australia.

Webber, L.M. & Ison, R.L. (1994). Participatory rural appraisal design: conceptual and process issues. *Agricultural Systems* **47**, 107–31.

8 The graziers' story

Danielle Dignam and Philippa Major[26]

8.1 Introduction – 'we are the bush'

As graziers the whole idea of research seemed pretty remote to us. It was something an academic did for a degree, then shelved along with many other projects perhaps never to be used again. We were surprised when approached by the Sydney team, that anyone would be interested in what our thoughts were in regard to research projects and their results. After talking with them and other interested graziers, we realised that we actually did research ourselves, but called it experience or local knowledge.

Through this project we have learned to do more of our own research, we have become far more informed about the wool industry and we have gained the confidence to ask questions and have a say in the things we feel strongly about.

8.2 Five snapshots of life in the bush

To get an idea of what this research meant to us, it is helpful to have an impression of who we are and what it is like living out here in the bush. We all have stories but these are just a few of them.

(i) *Growing up out here*

My father's parents came from South of Broken Hill, around Coombah and when he was thirteen he left school and went out shearing. Then in the late 1940s, the Western Lands Commission started dividing up the large properties and to be eligible, you had to put your name in a ballot. Dad drew 'Floods Creek', which was part of the huge Kidman property 'Corona', and with my mother, a couple of horses and a truck he settled himself there. It was just the block, no house, no fences, and no dams. He didn't even know where the boundaries were.

He bought sixteen hundred sheep and built a couple of dams and some fences and a tiny stone hut that was only four walls and a roof. It was so small that when my brother and I came along, we had to live in a tent next to the

---

26 / The introduction was written by Sandy Bright, one of the grazier participants in the research. The remaining material was prepared by Danielle and Philippa based on the contributions of Sandy and Cleve Bright and Kim and Margot Cullen, who responded to the invitation to prepare a chapter based on their experience.

house. Mum was a hard-working woman too. She used to walk behind the sheep and do whatever needed doing. After they got married, they didn't go into town for nine months.

From there he built a bough wool shed by cutting down a few trees and putting a bit of netting over the top for shade. And instead of just shearing on the dirt like everyone else did, he went and pinched a whole lot of tar from Silverton where they happened to be building a road at the time. They brought it back and laid it all down, but it happened to be hot when they were shearing and all the wool got stuck in the tar, and birds were landing and getting stuck in it. It was a real mess.

Another five years later he started building the shearers' quarters, and we lived in those for a while until he had built a new woolshed. We stayed in them for a few years until he got enough money together, and then he built the house.

He's been through some hard times, my father. The day the house burnt down I was in town, and dad was the only one home. Apparently the combustion stove leaked out all over the floor and the whole house was gone in fifteen minutes. He was sleeping out on the verandah and all he could do was watch it go. Then the shearers' quarters burnt down, and one person was actually burnt to death in that one. We were away on holidays then, and coming back we heard on the wireless that it was my father who had died. We were going along the road with tears running down our faces, and along comes the bloody hearse. We pulled up and the Funeral Director told us it wasn't my father . . .

He then bought 'Mount Westwood', and in 1977 bought 'Joulnie' as well.

It was pretty difficult when we started running the place with him, because although he said we should make our own decisions, he had been running the property successfully for all those years, and who were we to say 'this is how it should be done?' So I think he realised that and decided to leave. The bush life was all he really knew, but now he has adjusted well to Adelaide life.

There are a few pioneers left out here who could tell stories you would find hard to believe. But those stories will all be lost soon.

(ii) *The pace of life*

When I was a kid everything was slower. There just didn't seem to be any reason to go any faster. They even shore sheep slower in those days. We are shearing seven or eight hundred sheep a day now. When dad used to do it and run the same amount of property, and probably make more money, he was only doing it at four hundred a day. I suppose we have to keep our costs down because everything is so much more expensive. The money goes so much quicker and you are in debt so much faster – it is all over and finished before you have even had a chance to think about it.

(iii) *The sense of community*

Life is really changing in the bush. We don't make our fun like we used to. When we first got married we had no TV or mod cons. We used to invite the neighbours over for cards and we would go there. It would be terrific fun. There would be tennis matches at different people's places, and they'd play cricket out on the flat – they would just roll down the mat and play. If you tried to organise that sort of thing now, people would say 'I'm too busy shearing' or whatever. The neighbours used to come and help mark lambs, and you would go and help them. Now the modern things have taken away the values of life. We don't have the get-togethers any more, and so we aren't as involved with our neighbours. We spend our time watching TV or videos, or going into town for entertainment. Broken Hill has become the centre where people socialise and a lot of that other contact seems to have stopped. We are also busier because we are running more stock. Years ago we could survive on four thousand. Now we are struggling with twenty thousand. Because we are buying all these mod cons, you've got to pay for it with something, so you have to run more stock, and spend more time working. So you have less time for socialising.

Things do seem to be changing though. There seems to be a feeling now that it is getting too boring just living within the four walls, and people are starting to say 'there's got to be something else'. But they've forgotten or lost contact with how to do some of the things that they might have done ten or fifteen years ago. At least the realisation is there. But it is also good to be able to turn on the news and find out what is happening in the world.

(iv) *Isolation*

We don't fit anywhere. I honestly think that as far as a lot of people are concerned we have just dropped off the edge of the earth. We just don't rate a mention. It's amazing some of the letters we get – people have got no idea of what goes on out here – no idea at all.

(v) *Women on the land*

When we first moved out here we were the youngest people here and the men wouldn't let the women go out in the paddock – it wasn't women's work. Then last year my leg was broken because a cow trod on it while I was working in the cattle yards and the fellow next door said 'that's what happens when you have women in the cattle yards'. A woman is no slower in the yards than a man.

I have to laugh when people ask me what on earth I do out here. A couple of years ago we got a truck bogged in the creek with all our wool on it. The creek was running, there were big black clouds overhead and we had no tractor to pull it out. So we got the truck out using long handle shovels. There I was digging away and I remarked to the truck driver, '... and my girlfriend in town wanted me to go to aerobic classes'.

8.3 The story of how and why we became involved
(i) *Our first impressions of CARR...*

We didn't quite know what to think when the CARR team first arrived. We started worrying when their letter of introduction arrived. It could have been written in another language for all we understood of it. We had no idea what 'technology transfer' was or what a 'community approach to research' was, and we thought CARR must have been some sort of motor car club. Nevertheless we agreed to them coming out and talking to us, mainly because we were fascinated to see what sort of people would write such a letter. We certainly had no idea what they wanted to talk to us about.

We realised they were academics as soon as they pulled up in their Department of Agriculture truck. It didn't do much for their credibility in our eyes, but at least it explained their foreign language. The little experience we'd had with academics gave us the impression that they thought they knew a hell of a lot more than us and felt they could push their ideas onto us just because they had been to University and we hadn't. So we immediately thought 'what are these smart people trying to tell us and what is it going to do for us?' We wondered if it was going to be another one of those things where you answer a hundred questions but never see any results.

(ii) *Our experiences of research...*

Some of us had been involved in a research project which was run by a fellow from the Department of Agriculture, and it greatly coloured our view of research, particularly that funded by the AWC (Australian Wool Corporation).

This fellow came out a couple of years ago with a research proposal to look at stocking rates on particular pastures in our area. It required the use of some of our land and stock over a period of two years while he gained data. He agreed to give us the results of what he was finding in each of our paddocks. He gave us the impression that his research would be of great benefit and we were enthusiastic to be part of it because we thought 'here is someone who is prepared to do the scientific work to find out how we can manage our pastures better'. We felt that the results would have real practical benefit to our management because they were being generated from our properties.

He spent a day at each of our places, finding out about our current stocking procedures, looking at all our previous records and filling in forms about our management details. We also gave up the use of paddocks and sheep for the two year period. This was on the understanding that when he started to get results in, he would give them to us. But this didn't happen. After a while we started ringing him up to find out where the results were, and every time he would say he was still collating the figures and they would be ready in a couple of weeks. We felt that we had been deliberately lied to about the whole thing. We had become involved, given up our time and our land, and had received nothing in return.

The Wool Corporation and government departments aren't the only ones who don't consult us about our research needs either. It amazes us that we have a huge research station called Fowlers Gap in the middle of our area but they don't ask us what we want researched and don't tell us what they are doing. They just seem to research what they want. The only way we find out what research they are doing is if we happen to get into conversation with someone who works there. When we do find out about it, it often doesn't have any relevance to us anyway. At one stage they were doing trials on sheep and had them in really small paddocks. How can the research that comes out of one sheep in a small paddock have value to our management when we have huge paddocks with thousands of sheep?

Over a cup of tea the CARR team tried to explain their research proposal, but all we could grasp was that they were being funded by the AWC to research research, which didn't make much sense to us. While we didn't understand what they were trying to do however, they seemed to be different to the researchers and academics we had come across. It was nice for a change to be listened to rather than talked at, but we couldn't help wondering why anybody would come all the way from Sydney to have a cup of tea and find out what we liked about living here. Nevertheless we enjoyed telling them a bit about our histories and the issues we had, and they seemed genuinely interested in what we had to say. From that first night when they came to dinner, we felt we had some new friends.

(iii) *Going to the first workshop . . .*

After coming out to our properties individually, the team held a workshop for all the graziers before they left. Normally we'd think it would be a waste of time, but for some reason this time we had a strong feeling that something was genuinely going to happen, and we didn't want to miss out. However we couldn't help wondering what we were getting ourselves into. It was March and at least forty degrees in the shade. We couldn't believe it when we arrived and saw almost twenty other graziers there. It would have to be something pretty good to get so many people out in such heat.

While we sat around having a drink it was clear that nearly everyone was feeling optimistic. The team stood up and presented the issues which they had drawn from our interviews on big pieces of paper. We couldn't believe they had gained such a good understanding of us in such a short time. They tried to explain that they wanted us to decide which issues were most important and then help us research them, but they weren't going to do it for us and they didn't have the answers. A lot of people became very frustrated, saying things like 'If you don't know what you want to do, how on earth can we be expected to know?' Sometimes you have your problems for so long that you forget you've got them and you start accepting them as part of life. Your children are turning into delinquents, but you don't stop to think that it might be due to the problems in the education system. Others wanted to know what effect the team could have on the high taxes and the low commodity prices since they were the real issues. 'Did they have connections in Canberra so that something could be done about the things that really mattered?' There was a lot of misunderstanding and a lot of people who simply wanted someone to come along and solve all their problems.

While many of us were frustrated and confused, we were enthusiastic about being given the opportunity and the assistance to take our own action and become researchers. The idea of us researching our most important issues made a lot of sense. At that time we were facing a crisis with our wool prices and felt helpless to do anything about it.

(iv) *Our concerns about the wool industry...*

One of our main concerns about the wool industry before the CARR project came along was that we as middle micron wool producers were not being given a say in how our industry was run. It has always seemed to us that the super-fine wool growers have a lot more influence than other growers. For some reason they are the main group of grower representatives on the Board of the AWC which makes decisions about the industry. So they are supposedly representing the whole Australian Wool Clip. But in fact they aren't representing us at all – they represent their own interests. We produce seventy percent of the wool clip and we have no say whatsoever. While we realised that it was up to us to do something, over the years we have felt more and more helpless and have come to expect someone else to go and do it for us.

What we particularly wanted to be represented on was the issue of wool marketing. A few years ago fine wools weren't doing any better than broader or medium wools, but the Corporation started a big promotional campaign where they were just promoting the fine wools instead of concentrating on the broad spectrum – 'Cool Wool' they called it. We were still being paid enough for our middle-micron wool to survive on, although the gap between the prices we were getting and the fine wool was growing all the time. But as soon as the floor price collapsed we all became unviable. The stockpile was growing and our type of wool just wasn't selling at all, and all we kept reading was that superfine was skyrocketing.

The other major problem we could see with the sort of promotion they were doing of fine wool and its products was that it was turning wool into a luxury fibre. It was putting wool out of the price range of most buyers. All that superfine machine washable wool is just fantastic if that is what you're buying, but it has meant that many of us now can't afford to buy a woollen jumper. Recently a couple of us were in the Civic Centre in Broken Hill and there was a big wool exhibition going on. It was a big show, and there were about half a dozen other growers there, so we sidled up to one of the 'heavies' and said 'Okay, why can't we afford to buy what we produce?' He really didn't know what to say to us. All we wanted to know was why they seem to promote what we can't afford to buy. Why not promote what most of us produce?

Given the situation that wool was in, we strongly believed that they should be promoting wool in general – including middle micron wool, to increase the market and raise the price, rather than just promoting the kind of wool you can only wear in summer and which costs more than most people could afford. We believe we produce a pretty good product out here pretty efficiently, but our current returns are borderline. We are receiving $15 for sheep that require $16 worth of costs to get to that point. It is ridiculous. We felt a strong need to increase the demand for our wool through better promotion.

(v) *Agreeing to be involved...*

With our strong feelings about not getting a say in the wool industry at that time and the falling wool prices, we agreed to be involved in the CARR project. It certainly seemed to be different to the research we had experienced before and we believed we could gain something from it. We decided to focus our attention on investigating how we could better market our middle micron wool. Two years ago that topic wouldn't have been of interest to anyone, because everyone was making plenty of money and no one cared who was promoting it. Everyone thought the AWC was doing a terrific job. We would have had a different topic more relevant to the time. At that stage, however, the viability of our properties was being threatened by falling wool prices and we wanted to do something about it.

While we were enthusiastic about the possibilities, however, we were a bit worried about how we were actually going to do this research. We could grow the wool, but we weren't promoters and we certainly weren't scientists. But nobody else was going to do it for us.

8.4 The story of our research – 'the thirst for knowledge'

(i) *Where to start?*

It soon became clear that deciding on the focus for our research was the easy part. The hard part was deciding where to start. We felt like we were at the foot of Mount Everest with no equipment, contemplating how we were going to get to the top. There were just so many things we needed to know. We hardly even knew what happened to our wool after it left the front gate. It became clear that there was a lot we needed to learn if we were going to have any influence.

(ii) *Fax from the AWC...*

One of the first things we realised was that if we were going to learn anything, we would have to ask a lot of questions. We spent many hours thinking of all the things we needed to know about marketing middle micron wool, all the while growing increasingly aware of how little we knew about our own industry. We finally faxed four pages of questions to the AWC in Melbourne, feeling pessimistic about our chances of getting a response. Would they take us seriously? Would they laugh at our questions? Would our fax get lost in the vast bureaucracy of the AWC?

We couldn't believe it when we received a reply within days. We were even more amazed to find that the person who replied had taken the time to write seven pages of answers. It seemed like a breakthrough. It was the first time we had been able to think of the wool corporation as being made up of real people who were interested in what we had to say.

That fax gave us real momentum. We had taken the first step and had not been laughed at, we had answers to questions we had wondered about for years, and we had a lot more questions we wanted to ask.

(iii) *The difficulties of finding information...*

It certainly hasn't always been that easy finding information and answers to our questions. In fact often it has been a real battle to find out where our wool fits into the picture and what they are doing to promote it. For some reason a lot of information generated by organisations like the Wool Corporation is not at all readily accessible to graziers and isn't in a form which we can understand. The AWC is supposedly there to assist graziers but to our frustration, we found that even their definitions of certain things were not the same as the growers'.

At one meeting we decided that it would be interesting to find out the average micron of wool produced in our region. We sent off this request to the AWC thinking they would have the information at their fingertips. We should have known it would not be that easy. The statistics came back with all the wool lumped together – the lambswool, fleecewool and everything, so the regional micron average showed up much lower than it actually was. We tried to get more specific results, and received another lot of statistics broken into merino and crossbred. This seemed very strange, so we contacted the AWC again and found out to our amazement that any wool over 24.6 microns is classified as crossbred. We couldn't believe it! The bulk of the wool produced in the area is over 24.6, and unless someone was smuggling crossbred sheep onto our properties, as far as we were concerned we only ran merinos. So after two generations of calling our wool middle micron we discovered that as far as the Wool Corporation was concerned most of us weren't middle micron wool producers at all. It took us weeks and weeks of discussion and faxes to find that out.

At one of our workshops we had some representatives of the Wool Research and Development Corporation come and see what we were doing and we took the opportunity to ask them a few questions about wool marketing. We had found out previously that it takes four or five years to reduce fleecewool by one or two microns, so we asked them whether there was any foreseeable benefit for graziers in doing this. Their reply was that we needed to look at the proportion of the market we occupy, look at the difference between our net results, and compare this with the returns for the lower micron wools. They also told us we would need to find out who was buying our wool and what they were using it for. We were stunned. It had taken us six months to just find out what we were producing, let alone who was buying it.

The more information we found, the less we felt we knew and the more complex the whole area of wool marketing seemed.

(iv) *Visiting the AWC* . . .

The whole way we thought about the AWC changed when we went down to Wool House in Melbourne. As a producer you tend to think that people in those sort of positions can be a bit stiff in the upper lip and bogged down in the bureaucracy. So we organised meetings with various marketing and research officers feeling extremely nervous about how they would receive us. We wondered if they would be at all interested in what we had to say and whether we would know what to ask them.

When we met them they were totally different to what we expected. Rather than the bureaucrats in suits who did their jobs without caring, which we expected, they were open-minded and flexible in their ideas and extremely concerned about the wool industry despite being uncertain about the future of their jobs. They bent over backwards to help us, and everyone we spoke to was more than happy to answer our questions. They were surprised and concerned that wool growers knew so little about what the AWC did and one fellow even said he would like to come up and visit the group to see how we operated and find out our ideas on the wool industry. We thought that was just fantastic. It was great to see blokes like them in professional positions and us as growers communicating like that, because that is where the breakdown has always existed.

What struck us was that there was so much more information available from talking to the people in person rather than over the phone. They showed us how they were trying to find different ways to utilise the type of wool we produce and were about to launch their new advertising campaign encouraging all Australians to go out and each buy a pair of woollen socks. Previously we would say 'why don't they do this and why don't they do that' but after talking to them we understood why they couldn't do certain things. If only the growers were told more about what was happening at the AWC, they wouldn't spend so much time whingeing at them. As we learned, they really are doing a lot for our wool.

We returned from that trip with great enthusiasm and motivation to find out more. We had gained a much broader outlook on what was actually happening in the Corporation, and felt that they might have learned something about us in the process. We were confident that we could approach any of the people there to find out anything we needed to know in the future – a real development from our nervousness when we arrived. And as always, we had many, many more questions. We couldn't wait to get the rest of the group down.

(v) *Visiting the Mills . . .*

Our next trip was to the Riverina Wool Combing Mills in Wagga Wagga.

We were starting to feel like intrepid explorers – no mountain was too high, no challenge too great. Our lives were taking on a new dimension. Instead of classing our wool without thinking, we would wonder how the mills would want it classed, whether they preferred owner-classers or professional classers, and whether they went by the AWC's Code of Practices for classing. Our heads were constantly full of questions, and of wondering who we could ask. We felt we were starting to get somewhere with the whole wool marketing puzzle, which only a couple of months ago had seemed so complex, and it gave us the momentum to keep going.

But while it is one thing to have the questions, it is another to have the confidence to go and ask someone for the answers. In the past we spent a lot of time sitting back and complaining that people weren't doing this and that for us. With the experience we were gaining from our research, we were starting to feel more comfortable and a bit less shy about asking people if we could go and have a look at something or ask questions. As we gained momentum we felt more and more like we knew what we were talking about.

After hearing other growers talk about their experiences at French Mills and the International Wool Secretariat, we decided that direct selling to mills held some possibility as a way of breaking through the slow and expensive auction process – and we wanted to see it first hand. So a trip was organised to the Wagga Mill, and we set off with a hundred questions and the usual nervous excitement.

We were welcomed with open arms by the Wagga Mill people. Nothing seemed to be too much trouble – they took the time to show us the whole process and went out of their way to answer our questions. We got a very different view from seeing a set-up where they buy most of the wool direct from growers as compared to the brokers we were used to. They were far more specific about what they wanted from the growers, and seemed very keen to buy our pastoral wool because it had better length and strength than other wools and it particularly suited their mill.

It was an important trip for us because for the first time we realised that we could sell our wool direct to a mill with the assurance that we would get a good price. In the past private buyers have tried to take advantage of the grazier because that was how they made their money. All they did was pack the wool and put it in the auction system and so the money he made was at the grower's expense. Selling to a mill like Wagga is different because they are not selling back into the auction system. They are value adding and then making money from that. Therefore they are not making money at the grazier's expense. Because they buy wool direct it also meant that we could save on selling costs, which can be as high as $16 000 in a year. Unlike the auction system, where we have to get our wool in a month before the sale and then wait three weeks afterwards to get paid, this way the grower gets his money on the day they measure the wool in the shed. We were extremely motivated about the possibilities that had been opened up to us, and rushed back home to tell the other growers.

When a buyer from Michell's, another one of the big mills, came out soon after with an interest in buying our wool, we jumped at the chance to visit another mill. We wanted to build on what we had learned, finding out what sort of wools they were after and why, and comparing the operation of the two mills. Once again the staff were extremely friendly and helpful. They opened the machine up and put it on slow motion so that we could see how it worked, which was just amazing. We were particularly impressed with the blending process. They mix the wool in every way possible with the result being a perfectly mixed sliver, so no matter where you pick a piece, it is exactly the same as any other place. It made us wonder why they specify they want a certain type of wool if it all gets mixed up anyway.

On top of the advantage of selling direct, we learned that Michell's have a system where the growers can sell their wool forward to capitalise on good wool prices. You shear it, and sell it using last year's measurements and today's price. When it comes in they do the tests on it and then they alter the price depending on the difference between this year's wool and last year's. So if you think the wool price is going to go down, you can sell it today, deliver it in February and be better off. If you think it is going to go up, you can sell it in February and get February's prices. That is great for us. And they guarantee the price once you agree.

What we found out down there actually led us to sell some of our wool through them and we were really happy with the results. It has given us an alternative – a different way of looking at marketing our own wool. One of the graziers in our area even took his whole shearing team to Michells, because he wanted his shearers to do the right thing by him in terms of the Mill's specifications.

After visiting the Mills, we felt very strongly that everyone should know what happens after the wool leaves the property. When you have the full picture it gives you a new perspective to what you are actually doing on your own property and why you are doing it. We feel like we have learned so much in the last 12 months that we must have been going around with blinkers on before that. Obviously we have always cared about what happens to our wool, but we have just never known how to go about finding anything. One chap in the wool corporation even said that we would be in the top five percent of growers who know what is going on in the industry. That has got to make you feel pretty good doesn't it? But it also has a down side to it, because sometimes the more you know, the more angry you become. For example we were furious to learn that there is only about $2 or $3 worth of wool in a $300–$400 suit. The rest of the cost is absorbed by the processing side and of course there is a huge markup because it is a luxury item. That really hurts us because we are not getting the money and we are the ones going under. Even if we got a dollar extra it would make an enormous difference.

At least knowing what is going on puts us in a better position to do something about it. Now every time we go through the Stock Journal we say 'well that's relevant to what we are doing', and often we will follow it up by writing away for information or whatever. Recently we saw an advertisement in the paper for a forum by Michell's on 'Market Demands and How the Producer Can Meet Them'. We thought 'that would be pretty relevant to us', so we went down to Adelaide to attend. It is little things like that which we are doing now which we never would have done before.

8.5 Stories for the future
(i) *Achievements*

We learned important things about the industry we are involved in and made valuable contacts through our research.

Times are changing for us as graziers, and with world markets and government policy seriously affecting our livelihood, we are realising that we can't afford to sit back any longer and let others make our decisions for us. It is becoming clear that graziers have to become more aware of what is happening in our industry, take our own action and be able to stand up and tell people what we think.

Being involved in the CARR project has helped us feel like we can actually do that. That doesn't mean that we have all turned into politicians. We are graziers first and that will always be our priority. But now that we have learned so much about the wool industry and can go on learning, we can prove we aren't idiots sitting out here in the middle of nowhere. We are not ignorant and we know what is going on in the world. And we now know that we have a voice which can be counted if we want to make our voice heard.

We developed the confidence in ourselves and our abilities to be able to approach people in important positions and ask questions, even when we are fairly ignorant about the subject areas.

Overseas one of the graziers spoke to the boss of the Mill in France. Who would have thought twelve months ago we could have done that? We would be too worried we would make a fool of ourselves because we wouldn't have known what to ask.

The CARR researchers proved to us that our input was not only listened to and acknowledged but utilised and followed through. This was completely different to the research we had experienced before.

When the CARR team first came out, they told us that there were two kinds of research: research *on* people and research *with* people and that they wanted to research *with* us. We didn't know what they were talking about then, but now after a year of researching with them, it seems like the only way to make the results of research useful. With this project, we knew it would be relevant because we would be doing it ourselves. If someone else does the research, more than likely we will never look at it.

This has led us to question the point of doing research that is not useful to anyone. The WRDC gave us a big book on all the research that they funded, and we were shocked. We had absolutely no idea that so much research was done and that there were virtually never any graziers invited to contribute ideas on what research should be funded. We had certainly never seen the results of any of those projects. It occurred to us that there must be a hell of a lot of research done that nobody wants to know anything about. Somebody wakes up one morning and asks themself why the sun comes up and the next thing you know there's a research project for two years on that. It is a huge waste of their time and our money.

It has changed the way we look at research. The next time someone rings us up or knocks on our door it won't be just a matter of inviting them in, filling out a form and never hearing from them again. We will insist that there will have to be a bit more feedback than that. It has been proved to us by the CARR team that people can follow things through if they choose to.

Part of that was realising that we actually do our own research. When the team came out and told us they wanted to research with us, our immediate reaction was that we were graziers, not scientists. To us research was something that academics did and we had no time for. When they asked us how we worked out how many sheep to put in each paddock, we told them it was through experience and local knowledge. But the more we got involved with research, the more we realised that as graziers we do a lot of our own research by just doing something and seeing that there's got to be a better way – only we had never thought about it as research.

(ii) *Limitations of this approach*

While we have learned a great deal from the experience, we have got to the point where there is a limit to what we can do. We will keep asking questions of different people but it just takes so long to get anywhere and time is something we don't have a lot of. To keep the CARR project going at the rate it was, we would need someone who could work on it full time to help follow up our questions. The next step for us is to go overseas and talk to buyers to find new markets for ourselves, but how are we going to do that? We don't have the time or the money.

The main consideration is that we are graziers first and that will always be first as far as we are concerned. Before anything else we have to deal with the practical issues affecting us and that does not always leave time for going to conferences or visiting wool mills or reading up on the new developments in wool marketing. When we are in drought there is no time or energy to do anything but work on staying viable. When there is a flystrike outbreak, it is a full-time job to control them and minimise stock loss. For the couple of months a year when we are shearing and crutching that is the only thing we think about. So research is great, but it is not always high on our priority list of things to do. And as we said before, wool marketing will not always be the issue we are most concerned about. Next year it might be education issues, or woody weeds, or feral goat control.

The other limitation to the CARR approach as far as we see it is that achieving something as a community relies on everyone wanting to be part of it – that's your biggest hurdle. With this project it has been difficult to get a lot of people committed to working on it, but those of us who have been involved have got a hell of a lot out of it. It's all very well to talk about these community things, but unless it means dollars in the pocket or some other important benefit, people just won't get involved.

The question is what does it take for people to get involved in something? One of the things that was brought up at the Wagga Mills was that they would like to come out here and buy about half a dozen growers' wool clips as a community venture. But once again that relies on people being prepared to take a chance and try something new and many out here wouldn't. When the Army came out to the area to do their military exercises, everyone for miles around got together at a meeting to jump up and down and protest. People were spilling out the doors – we have never seen so many graziers together out here in our lives. But people won't go to something which they think may not directly affect them. For example, we don't go to the Bushfire meetings because we think 'well there aren't any fires around and it is a long way to get over there so we won't go'. But we know damn well that if there is a fire we'll be up and at it. So perhaps if an alternative comes up where we can sell our wool or promote it a bit better so that we can save the wool growers a bit of money, they would all jump on. That is what we have got to find.

(iii) *Possibilities for the future*

While the CARR group will not keep going in its present form, as an idea or a concept, perhaps we can apply it to something else. The important thing is that the graziers get a say in what's going to happen out here – we are the ones that have to live out here for the rest of our lives, and we can't let them dictate the terms. There is a big interest in Landcare at the moment because people are aware of the fact that if we don't get together and improve the land then someone else will impose it on us, and more than likely it will be someone who doesn't know what they are talking about. Landcare is about working with community groups on issues that we are concerned about, similar to the CARR project. Perhaps it is an opportunity to continue our research.

If it wasn't for this project we would still be thinking all the questions but not asking anybody.

Being involved in a project like this is tremendous actually. I feel part of it, I feel part of the team and I feel important, and that's really good. I have really enjoyed all of it. I get a kick out of it. I can see the benefits of being involved. Going out and learning things and finding out for ourselves has made it all really worthwhile.

Our little steps have turned out to be monumental leaps.

Part IV
Limitations and Possibilities for Research and Development Design

It would have been possible to have the graziers conclude this book (Chapter 8); to have the final word. This would have been fitting. The question of who speaks and who has the final word... before moving on, is an important one. In a sense though they have, because they have said what they wanted to say. It is clear that for these graziers this project changed their lives. Perhaps not profoundly, but meaningfully. A context was created in which they were able to respond – they were able to pursue their enthusiasms with outcomes meaningful to them. They now bring their new understandings to current conversations or projects. So do we. We have all moved on to new conversations, new dances, yet our experiences as part of this project inform our actions in diverse ways.

The rationale, therefore, for our final chapter is that because our overall research project is not unlike a first draft (we were breaking new ground) we feel some enthusiasm to address the following questions: What are our reflections on the overall experience, or put another way, what phenomena did we experience for which we now seek explanations? What do we do now and how is this informed by this project? This is a return to the model of doing science which we addressed in Chapter 6. Our experience of this project convinces us that an alternative way of doing R&D, a complementary way, is possible. We explore our fantasies for a different R&D system in which contexts are created where there is greater capacity for people to respond.

Our research, and the description of it in this book, is not unlike the approach to doing history related by the Australian historian, Manning Clark:

There was never a grand design or an overall plan... The movement from chaos to design was never planned or seen. It just happened... There was a first draft of roughly two hundred thousand words for each volume. That was generally hopeless, useless, an embarrassment. There was a second draft, again laboriously deciphered by the long-suffering research assistant, which had some resemblance to what I had dreamed might one day appear on the page. There was a third draft which incorporated some parts of the second draft – both being approximately two hundred thousand words. Then there was the long-drawn-out fiddle and mucking around with the third draft, much of which was more a fourth draft than a fiddle with the third, until the whole seemed to me to have the design and tone for which I was aiming. My model was the sonata form in music – the statement of the theme, the development of the theme, the recapitulation and the coda.

– Manning Clark from An Historian's Apprenticeship, 1992.

As with Manning Clark, this is just the first draft of our story. It is the first attempt at doing something different and we have made many mistakes and learned a great deal. What is important is that our story of the CARR project is not a methodology to convert anybody. What it has shown us is that people will go on doing what they want to do. We are not setting up something that will solve their group problems or the team functioning, or pledging this to be the way to go in any sense. It is another example of an outside influence that doesn't change people's hearts and minds in any predetermined manner.

It is easy to fall into the trap of seeking to convince people that this is the way to go. What we want to do is record what it is we did, and say that we feel it has been highly successful in the terms that we set up. It is not a recipe for going out and working with the community. Nor is it a research procedure to replace other procedures of the biological, physical or social sciences – it is just another approach. It lies alongside those and is complementary. It is very good – if you want to do it this way. It will not be good if you want to influence people. However, most people do want to influence people. They want to advise them of something they can do better. It is very rare for people like ourselves not to want to influence anyone.

We have tried to capture the uniqueness of what we were doing because at the time it was so different to anything else we had encountered. This is what is at the heart of this book. It has been important that we emphasise this because immediately people will try and interpret what we have written as another influence exercise. Our hypothesis was that if we were there with the graziers in a supporting role, then they would do with more vigour what they wanted to do anyway. Whereas, helping in our culture is usually understood as someone having new or better knowledge or skills, and you go to them for help because you hope they will shed new insight or whatever. People would say 'What have you got?' We would say 'We just want to listen to you, we just want to support you to do what you want to do. We have nothing to offer other than a real willingness to be in conversation with you about what you are doing'.

Our efforts in researching an alternative approach to R&D were triggered by experiences similar to those which led Maturana (1991) to claim that: 'In our modern Western culture we speak of science and technology as sources of human well-being. However, usually it is not human well-being that moves us to value science and technology, but rather the possibilities of domination, of control over nature, and of unlimited wealth that they seem to offer. . . . We . . . speak of progress in science and technology in terms of domination and control, and not in terms of understanding and responsible coexistence'. 'What science and the training to be a scientist does not provide us with is wisdom. . . . Wisdom breeds in the respect for the others, in the recognition that power arises through submission and loss of dignity, in the recognition that love is the emotion that constitutes social coexistence, honesty and trustfulness and in the recognition that the world that we live in is always, and unavoidably so, our doing.'

Our research outcomes can be seen as a means to respond to Maturana's claims if, as a R&D practitioner, one is inclined to do so. In agreeing with Maturana's claims we are not becoming sentimental but conceptualising a second-order R&D system as a social system. To do this requires an explanation of 'social'. Maturana describes love as 'the domain of those relational behaviors through which the other arises as a legitimate other in coexistence with oneself'. From this understanding of love, arises the conditions which Maturana accepts as a description of the 'social': 'a couple arises as a simple unity (as a totality) in the moment in which two persons begin to move and behave with respect to each other as a couple, and it exists as such in the domain of social declarations where couples take place as civic entities'. Thus following Maturana and Verden-Zoller (in press), we would describe second-order R&D systems as social systems in which the others, the stakeholders, arise as legitimate others in relation to oneself (the researcher or researchers) and each other.

References

Clark, M. (1992). *An Historian's Apprentice*. Mebourne University Press, Carlton, Victoria.
Maturana, H. (1991). Response to Berman's Critique of The Tree of Knowledge. *Journal of Humanistic Psychology*, **31**, 88–97.
Maturana, H. and Verden-Zoller, G. (1999). *The Biology of Love* (in preparation).

9 Designing R&D systems for mutual benefit
David B. Russell and Raymond L. Ison

9.1 Introduction

The aim of this final chapter is to provide the reader/practitioner with the requisite fundamentals to fashion a researching/learning/action system designed to meet the needs of different and often conflicting stakeholders. It is justifiably an action system because it has its foundations embedded in the world of experience. It is a learning system because it has the built-in capacity for reflection on experience and the wherewithall to recognise change or learning when it has occurred. It is a researching system because it offers contestable knowledge, knowledge which is open to robust and critical appraisal.

The insights that go to make up this design process are derived from the research project in the semi-arid rangelands of Australia. The application of these initiatives, however, would seem to be suited to a whole array of possibilities from organisational change in public or private enterprises, across tasks like team-building projects, to the development and delivery of educational programmes. The strength of the proposed design approach is that it offers a transparent way of working with conflicting interests and seemingly contradictory goals.

The essential characteristic of the proposed 'system', the one that when it is not present would mean that the system was no longer in existence, is that the accepted needs of the key stakeholders are being addressed and adequately met (as judged by the stakeholders themselves). What this means is that there needs to be sufficient plasticity within the system for specific relationships to be formed and dissolved so as to ensure that the *sine qua non* of the system, its capacity to satisfy the agreed-upon needs of the critical stakeholders, is maintained.

In specifying the constituent elements of this R&D system we will return to notions of first-order and second-order systems that have proved to be so useful throughout the study. In specifying the nature of the constituent elements we will use as a heuristic the following question: How would an outsider know that she/he had encountered an effectively functioning second-order system? What would be the markers that would be open to public experience and would be agreed upon by a specific community as being the essential ones?

A complementary set of behaviours, ones that represent the requisite skills necessary to bring forth the R&D system, will also be named. It is the view of the authors that the skills required to implement the sort of design process that we are talking about are akin to essential technologies, those social extensions of the person which allow certain experiences to come about.

9.2 The general factors that constitute a social system

The term 'system' is a convenient artefact often used to refer to a set of operations and relationships that have some specified outcome. It is when such systems are judged to be problematic or failing to achieve the desired outcome that they become the focus of a planned intervention. Such has been the case with rural development. The naming of the problem and the attributing of ownership to the problem often prove to be very contentious issues.

9.2.1 *Characteristics of a first-order R&D system and first-order problems*

The term 'first order' only has meaning in relation to a second-order system. When a set of operations and relationships are understood and described with little reference to how the system was brought into existence, its social and cultural context, then this is a first-order system. When the same set of relationships, events and operations refer to the 'how' they came into existence and 'what' they hoped to achieve (are self-referential), then it becomes a second-order system. First-order data describe the system as if it was an objective set of operations functioning independently of its historical and social creation. In other words, its existence as a useful artefact has been forgotten. Examples of such data would be descriptions of physical, biological, social, and psychological events where such descriptions are made without reference to the experience (past, present and anticipated future) of gathering the data.

9.3 Design principles for the development and evaluation of a R&D system

What follows is a four-stage strategy which incorporates both first- and second-order data collection. The strategy also integrates, in a manner congruent with the theory, the stages of evaluation and implementation. The R&D system has been conceptualised as a decision-making and action-taking system. Each stage is arranged as a table in which the critical tasks are described, the associated processes listed (be they first-order, second-order or a combination of the two), the requisite skills noted and finally, potential pitfalls are recognised. This articulation of the strategy is one of the outcomes of the research project in the semi-arid rangelands. While all the elements were present in the study, they were not organised in the systematic way that they are now being presented.

9.3.1 *Stage One*

Table 9.1
Bringing the system into existence (i.e. naming the system)

Tasks	First-order processes	Skills	Potential pitfalls
Agreeing on the essential participants (the key stakeholders)	Invite relevant parties to state their interest in a particular event/experience	Ability to identify parties with a particular stake in a outcome (e.g. resource providers; users of outcome; producers of outcome)	To equally involve stakeholder groups that historically have exercised little influence on how particular decisions are made
A 'system' is generated which has been determined by the main issues of concern to the key stakeholders	The system is determined by the 'problem' not the problem being determined by the system	Group process skills coupled with outcome-oriented skills	That preconceived ideas of what constitutes the 'problem' will hinder a reframing of what constitutes an actionable problem
Collection of sufficient empirical data so as to establish the existence of specified events/experiences	Generation of patterns of data over	Ability to recognise the key categories of data required and requisite skills to collect and quantify data	That easily quantifiable data will be judged as being superior to less easily quantifiable data (e.g. value statements; emotional responses)
Determining the boundaries of the system (conceptual; geographical, etc.)	To incorporate data from the bio-physical domain and the psycho-social domain in determining system boundaries	Ability to successfully invite participants to offer narrative data via social technologies (e.g. semi-structured interviews; focus groups)	To favour the generation of a dominant bio-physical system over a 'human activity system'

9.3.2 Stage Two

Table 9.2
Evaluating the effectiveness of the system as a vehicle to elicit useful understanding (and acceptance) of the social and cultural context

Tasks	Second-order processes	Skills	Potential pitfalls
Judgements on adequacy of data to contextual demands	Awareness of how data were generated and psychological and sociological driving forces at work (e.g. operation of dominant mythologies; historical underpinnings)	Ability to see different world-views as expressions of prior and differing life experiences	'Experts' and others with social status tending to impose their conceptual models and boundaries on other parties
Seek additional contextual data if necessary	Articulate the meaning-making linkage between first-order and second-order data: the latter giving meaning to the former	Ability to elicit contextual information and to appreciate the shaping function of dominant mythologies: how meaning is made by reference, often outside of awareness, to organising constructs such as institutional or cultural 'stories'	Desire to establish a hierarchy of knowledge 'types': one kind of knowledge being judged as superior (more useful) than any other type
Seek legitimation of match between an existing world-view and the history of how such a view was formed	Each individual is responsible for the world that she/he 'constructs' and each set of knowledge is valid for that person precisely because he/she has constructed it	Ability to work with a multiverse of world-views rather than aspiring for a common or universal view	That the researcher(s) will subtly try to influence the proceedings by asserting a dominant position representing their own point of view
Achieve 'two-way' conversation, or 'dialogue' in which individuals speak from their respective positions	Each expression to be accepted as a contribution of value to the eventual outcome (an outcome which is yet to be named)	To actively listen and respect (but not necessarily agree) with others. Skills to confidently present one's position	Some people are unable to accept that there may be different 'truths' representing different world-views

9.3.3 *Stage Three*

Table 9.3
Generation of a joint decision-making process (a problem-determined system) involving all key stakeholders

Tasks	Processes	Skills	Potential pitfalls
All participants (stakeholders) are encouraged to fully address their concerns and aspirations	Ambiguity and uncertainty reflect the non-absolutist understandings associated with every position	To reflect back to the participants how each position has an 'appropriateness' for a specified intellectual domain. Outside of that domain appropriateness diminishes rapidly	Matters which can be held as 'certain' in one domain can be generalised across other domains
Respective concerns and aspirations are mirrored back to participants showing understanding of respective position position	A publicly sanctioned reflexive process allows for both confirmation and public acknowledgment	Facilitation skills sufficient to reflect what has been contributed, and how it has been said, without introducing any new material or altering the emotional milieu	People not recognising and/or accepting their own blind spots
The 'problem', and thus a desirable outcome, is named	The problem is within the action domain of this group. This group has 'ownership' of the problem and of the eventual outcome	Skills of analysis and synthesis such that the nominated problem expresses some of the key needs of the stakeholders	That the responsibility for actioning the problem will be projected to parties outside of the task group
The decision (not necessarily agreed with by all) is made	The agreement is that all parties have been able to present their positions in a fair and full manner. Acceptance to proceed is not contingent on full agreement on the final position	That an intellectual and emotional climate is achieved in which all participants can see the merits of differing points of view and are able to 'let go' of preferred positions	That a stakeholder abandons the decision-making process rather than be seen to be compromising

9.3.4 Stage Four

Table 9.4
Evaluating the effectiveness of the decisions made (i.e. how has the action taken been judged by stakeholders?)

Tasks	Processes	Skills	Potential pitfalls
Collective judgements of how well the generated problem represented key needs of all stakeholders	This is a measure of internal effectiveness of the process and of subsequent commitment to the implementation of the decision	Ability to listen openly to participants' 'second thoughts' without showing excessive defensiveness	Risk of jeopardising the whole process because the outcomes were judged as being less than perfect
Assessment of increased readiness to address, in a similar manner, other needs and concerns	A second-order system coupled with a first-order system facilitates learning-to-learn by the participants and, increasingly, is embedded in the culture of the organisations	Ability to demonstrate the second-order outcomes and to present them as reusable building blocks	A tendency to disparage second-order outcomes vis-a-vis first-order ones
Estimate how transparent (open to public scrutiny) has been the decision-making process	a transparent process allows participants to accept a decision (because the process has been experienced as being fair and equitable) even when they do not fully agree with it	Ability to balance those who bring with them institutional and/or social 'power' with those traditionally less endowed	That a climate of mutual acceptance cannot be achieved: where differential 'power' has not been accepted
Evaluate the ease of implementation of the decisons made	Organisations tend to conserve their status quo, especially the desire to maintain the patterning of key relationships	Skills to articulate the structural variables, both constraints and enabling factors, which influence implementation	An institution might find it preferable to shift the entire debate to a totally different arena rather than implementing an 'upsetting' decision

Using the four-stage strategy as a template, it would be possible to critically appraise the procedures used in the rangelands' study. The current study did evaluate the first-order and second-order outcomes and the effectiveness of the processes used to achieve those outcomes, but did so in an informal manner. The above framework offers a more formal and complete research procedure and one in which the desired transparency, that openness to public scrutiny, is better achieved. The more transparent the system, the greater the possibility of usefully critiquing the activity.

The rangelands project was highly successful in that it accomplished targeted first-order objectives (e.g. demographic trends; technology audits – Ison & Russell, 1993; CARR, 1993a); specified second-order goals (e.g. graziers re-conceptualising themselves as R&D professionals – Chapter 8); and provided the raw material and emotional climate in which the R&D process itself could be scrutinised and modified so as to better match the multi-layered demands of its context. Certain second-order goals were not achieved – for example our work with research and advisory staff from NSW Agriculture did not lead to the establishment of a co-researching community (CARR, 1993b). Some of the reasons for this are found in Chapter 5.

9.4 Different levels of a system produce different data and suggest different changes

First-order data are collected out of a mind set of: Let's behave *as if* we could observe the specified events without taking into account the relationship that exists between these same events and the observers themselves. Or, let's behave *as if* this was an independent universe that we wish to observe, one which is unfolding before our eyes and one which we are progressively learning more about. When only first-order data are collected then only first-order problems will be agreed upon and only first-order change will be judged to be desirable. This mode of operating is especially important in areas where there is little or no disagreement about the validity of data, such as with physical and biological data, e.g. weather patterns over rural areas and specified production figures. The second-order data are totally different and are open to many forms of diverse interpretation. Second-order data are the outcome of the acknowledged interdependent relationship between the 'knower' and the 'known' through which both inevitably change over time and in which the relationship can be justifiably seen as bringing the data into existence. Examples of such data would be statements like 'One male in this family has always been a farmer!' And, 'We need better access to markets!' These examples are second-order not because they are of a qualitative nature (and not easily quantifiable) but because they arose out of a rapport between the researcher/observer and the person being interviewed, and/or, the interviewee reflecting on her/his own experience and seeing these reflections as constituting valuable data.

9.4.1 *Decision making and the increased capacity to meet conflicting needs*

First-order data, by itself, offers limited possibilities for effective decision-making when parties are in conflict over what is the most desirable action. When one set of first-order data is stacked up against another set there is no manner of reconciliation other than to agree to disagree or to make a claim of superiority of one set over the other such as when one hears . . . 'My data is more scientific than your data!'

When there is little or no first-order data and a preponderance of second-order data then the result is much the same. What we hear is 'I am right and you are wrong!' with any decision being made in favour of the party with the loudest political voice.

Understanding the two systems as being present to each other in a complementary fashion, gives both equal legitimacy that is denied one or the other when they are conceived of as being in opposition. Holding the two together, which is done when legitimacy is given to both, is not dissimilar to the Hegelian concept of a dialectic in which a tension arises between historically different understandings, experienced as being opposite ways of thinking, which is resolved by the proposal of a third thought which accommodates the best of both points of view.

Participants in the rangelands' study frequently spoke of being in possession of a broader range of options. This second-order outcome, by its nature self-referential, connotes an increase in decision-making capability in that there is a rational understanding that has changed (for the better) and which implies that making decisions is now a more satisfying experience. A further second-order change was the measurable increase in the networking of production information amongst graziers, which was attributed to the public acknowledgment of the validity of one's personal practice in a particular context. This new understanding was linked by participants to the public experience of seeing that any information was valid in the context in which it arose. Seeing how positions, including 'value' positions, were arrived at, over time, fostered an increased capacity to let go of fixed or rigidly held positions, all of which facilitated effective regional networking and the sharing of information for mutual benefit. Thus an observer is able to say that those engaged in this process are more *response-able*.

9.5 Individual enthusiasms and collective demands

The research described in this book clearly showed that encouraging individuals, by publicly validating their respective enthusiasms-for-action, resulted in the generation of sophisticated and effective R&D systems deigned to achieve the desired outcomes. The introduction of aspects of critical appraisal incorporating second-order processes, relating the data to the psychological and cultural factors of the context (past, present and anticipated future), generates second-order change – useful learning and useful 'learning about learning' – which is highly satisfying to the individual. What the research did not describe, because it was a finding by *omission* rather than by *commission*, was that 'systems generated problems' were not effectively addressed. Systems generated problems are those problems which do not arise out of the enthusiasms of the individuals but, rather, are products of first-order systems, systems which are perceived *as if* they had an objective existence. One such problem was the institutional need (in this case, the state Department of Agriculture) to conserve the system as it functioned at the time and, as a consequence, the structural barriers to any significant change that appeared when perturbations to the system were encouraged (see Chapters 3 and 5).

The final report of a NSW Agriculture project (NSW Agriculture, undated) conducted in the NSW western division concurrently with our own (1990–1994) is instructive on a number of points:

(i) Their findings which 'provide clear evidence of the extent to which market price fluctuations can outweigh biological productivity as a determinant of economic performance. While standards of on-property biological productivity must be maintained or improved where appropriate, pastoralists need to pay careful attention to product quality, particularly wool fibre diameter, and the price secured for their products'. They conclude 'attention to product quality and marketing . . . is essential . . .'. This was, of course the research issue nominated and pursued by the graziers in the CARR project, as described in Chapter 8.

(ii) It was acknowledged that in this project 'the department's agenda was always perceived by the cooperators to be paramount. The groups were put together by the department rather than forming spontaneously. Once formed, they saw the project as "ours" rather than "theirs". This style has provided good data to extend beyond the project to the pastoral industry in general . . . However, the lack of "self direction" may have limited its value to participants and the model is not considered satisfactory as a basis for delivery of Departmental services.' Implicit in this conclusion, it would seem to us, is a recognition of the differences between first- and second-order data and processes.

(iii) Their acknowledged difficulty in working in this part of the world despite their relative geographical proximity. This was also a constraint for us in our research and is one reason why, in our concluding report (Ison and Russell, 1993) it was envisaged that new structures for second-order R&D would be needed based on a network of self-organising 'problem-determined systems'. As we indicate in our framework (Table 9.1) individuals with special skills would be required to facilitate this on-going self-organisation (see Table I.1 in the introduction to Part I). It was our view that those individuals likely to be best able to carry out this function would be those who were enthusiastic to do so from local contexts and who were able to spend periods, of varying frequency and duration, outside these contexts for skill development and "supervision" as in some social work practice (a facilitated process of critical reflection).

The importance of acknowledging the existence of a system, one as it were, independent of our relationship to it, is not to be underestimated. From the pragmatic acceptance of such a system comes the notion of mutual obligation and responsibilities which are not anchored in enthusiasm-for-action. The emotions which underlie our responses, or usually lack of response, are more typically those of fear or the sense of being imposed upon or dictated to. The aim of holding together second-order change with first-order change in a dynamic yet gentle embrace, is that the eventual outcomes will be a combination of both responsible and response-*able* actions. It is only when we further develop our capacity, and the capacity of those with whom we collaborate, to achieve both levels of outcome that we can confidently speak of systemic change, change which in itself increases the range of options open to on-going change. Much of the capacity we speak of is conserved in on-going networks of conversation.

Institutions are slow to accept second-order processes because they offer a critique of how decisions are made as well as raising the complex question of what constitutes valid domains of knowledge for any particular conflictual situation. The same can be said for certain individuals who prefer to assert a manner of power/authority/self-righteousness over an issue, and/or, over other stakeholders in any process designed to achieve an alternative outcome. While no successful strategy for working with such individuals or institutional cultures can be guaranteed, the more transparent the process, the more likely that the majority of participants will be committed to implementing decisions that have been arrived at in a visibly fair and equitable manner.

Earlier chapters (see Chapters 2 and 6) referred to communication as 'a dance of understanding' in which the participants were encouraged to access their respective enthusiasms-for-action and put their 'desired action' before the public eye. The individual dances would not be the same, but they would be held together (and thus constitute a system) because of a mutual acceptance of the appropriateness (for each person) of each other's dance. They would constitute a social system in terms described in the introduction to this section. In the same vein we can extend the dance metaphor to include first-order and second-order processes and outcomes as each is appropriate for the context in which it arose. When the two move together (as a complementary pair) then there is a further expression of mutual acceptance and an elegance that quite befits a dance. The more explicit the choreography of the dance, the more transparent its underpinnings, the greater can be the acceptance of difference even to the point of the acceptance (which does not imply agreement) of opposing viewpoints. By proceeding in this manner it is hoped that both ecological sustainable development and cultural sustainable development will be achieved, *hand in hand!*

9.6 Concluding comments

On a more personal note, we have found the theory and practice of this second-order, systemic approach to be exciting and daunting all at the one time. Making our science self-reflexive and having the social, historical and intellectual contexts openly influencing the construction of our knowledge must be one of the most worthy pursuits available to humankind. As a task for science in today's world it seems to be especially relevant given our need to ask different sorts of questions. Research and development, like every technology, are a vehicle for the transformation of tradition. Being part of a tradition we cannot be objective observers of it. We can, however, let the potential for transformation guide our actions in creating and applying research and all that it entails. The transformation that we are most interested in and have a growing passion to further explore, is the continuing evolution of how we understand our surroundings and ourselves.

References

CARR (Community Approaches to Rangelands Research) (1993a). *Marketing of Middle Micron Wool. Researching with People on Issues that Make a Difference.* Monograph. University of Sydney & University of Western Sydney. 44pp.

CARR (Community Approaches to Rangelands Research) (1993b). *Institutions Responding to a Community Request: Wool Marketing Advisory Support Group . . . and Beyond.* Monograph. University of Sydney and University of Western Sydney. 48 pp.

Ison, R.L. and Russell, D.B. (1993) *A Collaborative Study to Develop Technology Transfer Systems More Suited to Semi-arid Rangelands with Emphasis on the NSW Western Division.* Final Report to Wool Research and Development Corporation, Project USY90. University of Sydney and University of Western Sydney. 32 pp.

NSW Agriculture (undated). *The Western Uplands Landsafe Management Project.* Final Report to Murray-Darling Basin Natural Resources Management Strategy (N061). 115 pp.

Appendix
Scientific explanations

One of the difficulties that is always present in our culture when one attempts to explain any experience, is that of realising when one has indeed explained what one wants to explain. With this difficulty in mind, let us say a few words about explanations in general, and about scientific explanations in particular.

We learn at home in our early childhood that an explanation is a particular kind of answer to a particular kind of question. Specifically, we learn that explanations entail the satisfaction of three conditions, namely:

1. that an explanation is an answer to a question about some particular experience, which explicitly or implicitly begs an explanation as an answer;
2. that in order for the answer to be an explanation, it must be accepted as such by the person (observer) who asks the question; and
3. that an explanation has the form of a process, that if it were allowed to operate, would, as a consequence, generate in the observer the experience to be explained.

We have called the process indicated in point (3) above, the generative mechanism. We wish to insist that a generative mechanism is not by itself an explanation of anything, and that a generative mechanism must be accepted as such by the observer who asks the question, to become in fact an explanation in the domain in which it is accepted. In these circumstances, an answer to a question that begs an explanation as an answer to be acknowledged as an explanation, must necessarily fulfil two conditions:

(a) it must have the form of a generative mechanism (we call this the formal condition); and
(b) it must satisfy any additional condition put in his or her listening by the observer (we call this the informal condition) to be accepted by him or her.

The informal condition is usually not made explicit, so different observers may listen at the same time for different informal conditions even in situations where they seem to be in basic agreement. Indeed, the informal condition defines the kind of explanation that an observer wants and accepts. In these circumstances, unless the informal condition is made explicit, it is not possible to know what is accepted when somebody accepts a particular generative mechanism as an explanation in answer to a question that begs an explanation as an answer. For this reason there are as many kinds of explanations as there are different kinds of informal conditions that can be put in the listening of the observer who wants to hear an explanation.

Scientific explanations are no different from explanations in general. What is particular to scientific explanations is the informal condition that gives them their peculiar character and specifies science as a cognitive domain. Furthermore, what is peculiar to scientists is that they have made their profession to be impeccable in the use of that informal condition for all their explanations. We call the informal condition that we scientists use in order to accept a particular generative mechanism as a scientific explanation, *the criterion of validation of scientific explanations.*

Put explicitly, this criterion consists of the coherent satisfaction of four operations to be performed by an observer, and one of which is the generative mechanism proposed. These four operations are the following:

1. The description of what an observer must do to have the experience that he or she wants to explain.
2. The proposition of a generative mechanism such that if it is allowed to operate, the result is that the observer has the experience to be explained.
3. The deduction from all the operational coherences entailed in **point 2**, of other experiences that an observer could have, and of what operations he or she should do to have them.
4. The realisation by an observer of the operations deduced in **point 3**, so that if he or she has the experiences also deduced in **point 3**, then, and only then, **point 2** becomes a scientific explanation.

Let us now make a few comments about the criterion of validation of scientific explanations and about what we scientists do with it.

- The observer is any person that can satisfy the criterion of validation of scientific explanations. So science is the club of those persons that use the criterion of validation of scientific explanations to explain what they explain, and any particular science is a particular domain of statements considered valid because they are validated through scientific explanations.
- Science does not require any assumption about an independent reality. Science does not explain an independent

reality, nor a world independent of what the observer does. Science explains the world of experiences of the observer by using the experiences of the observer.
- The aim of science is explaining and understanding, not prediction or effective action. If there is understanding effective action is possible within the domain of operational coherences proper to the domain of understanding. Scientific statements are valid in the domain defined by the scientific explanations that support them. If one is not aware of this, one may expect that science should allow for predictions of particular events that do not belong to the actual domain of validity of the scientific explanations that support them, and thus be mistaken.
- The criterion of validation of scientific explanations is the same criterion of validation that we use in daily life in a non systematic manner. The difference between science and daily life rests in the fact that a scientist is a person that lives under the passion of applying the criterion of validation of scientific explanations, is careful in not confusing domains when doing so, and is ready to abandon any accepted generative mechanism when such criterion is no longer satisfied. It is because science explains the human experience using human experience, that science transforms human life. We change the world we live as we live through science.
- The observer chooses the experience to be explained from his or her domain of experiences as a poetic act of free imagination[27]. The generative mechanism is also proposed in a poetic act, and it is specially designed by the observer using elements of his or her domain of experiences so that if allowed to operate, it will give rise to the experience that he or she wants to explain in the domain of his or her experiences.
- Once a particular generative mechanism has been validated as a scientific explanation, it is treated as an experience (that is, called fact) which can be henceforth used for the generation of new questions, or for use in the proposition of other ad hoc generative mechanisms in the explanation of other experiences.

27 / We define a poetic act as the detection of coherences in experiences in one domain and expressing them in another. This usually takes place in an unconscious manner which may be experienced as a revelation by the observer.

- Explanations are manners of human relations in the domain of consensual coordinations of consensual coordinations of behaviors; and as such they are not valid in themselves. Or, in other words, no generative mechanism is an explanation by itself, it requires to be accepted as such by an observer in the context of a question that begs an explanation as an answer'.

Source: Maturana and Verden-Zoller (1999) *The Biology of Love* (in preparation). Used with permission.

Glossary

action research	Action research is concerned with formulating and solving problems through a series of 'cycles' of plan, act, observe, and reflect. The aims are for individuals to learn by doing, and through experience, gain insight and understanding. This may lead to improvement in practice and alter existing situations and constraints. The espoused role of the researcher is that of participant–observer. In practice, however, the researcher often remains 'outside' the system being studied.
active listening	Listening in such a manner (verbal and non-verbal cues) that gives rise to the experience of being heard unconditionally. It is often a cathartic experience for the person being listened to and is emotionally demanding on the listener.
agreement	Agreement – that understanding has occurred – is dependent on the observation of a harmony of emotions (often metaphorical) underpinning a particular use of language which is fundamentally satisfying. Agreement (with understanding as its basis) is experienced as an emotional rapport which engenders trust and the confidence to recognise options and possible solutions to issues at hand. Agreements, when they only serve to coordinate our action, are often achieved at an emotional cost. Such an agreement can be at the expense of personal and social wellbeing and can also be counterproductive when any subsequent actions are either half-hearted or even destructive.
agricultural extension	The term 'extension' derives from the North American land grant university model where the goal was to extend knowledge from a centre of learning to those in need of this knowledge.
autonomy	The capacity of an organisationally closed system to self-generate, i.e. where a system's control over relations of production are wholly internal.
autopoiesis	This is the self-producing property of living organisms which is, itself, based on the self-referring process of cognition whereby the organism maintains itself in the world.
awareness	The conscious act of reflecting on one's tradition of understanding; epistemological awareness, we contend, is a desirable condition for the design of R&D systems.

blackbox A term arising from first-order cybernetics in which a system is conceptualised into which inputs are observed to lead, and outputs are observed to emerge, but in which the internal transformations are either not considered or not known.

cognition Refers to the activity of the nervous system. Contrary to the view that the nervous system picks up information from the environment and processes it so as to provide a one-to-one representation of the outside world in our brain, we see (following Maturana & Varela) that the nervous system is closed, without inputs or outputs and that the cognitive operation reflects only its own organisation. This implies that our interaction can never be instructive, i.e. unambiguous external signals. Rather, they consist of non-specific triggers which disturb the system but do not determine the nature of the response. Cognition is not an information-processing operation, but a constitutive mechanism of all living things.

communication In everyday language, 'communication' usually refers to the transfer of information as if such a process were indeed possible in anything other than a metaphorical manner. Rather than transferring information, we express our own version of reality that we each have constructed in the course of our daily living. Human communication needs to adequately account for the properties of the observer as well as the observed. It is to the mechanism of engagement that we must look to explain the dynamics of change occurring during human communication. The effectiveness of our communication will be dependent on the manner of our engagement.

conversation Conversation is the experience of understanding that is generated by the flow of emotions. Because the flow of our language and our emotions are so delicately interwoven, it follows that emotional matching is the precursor of semantic congruence. The meaning in the conversation will only match when the emotion matches. Whole cultures arise through networks of conversation leading to widespread agreement about concepts and values and a comfortable ability to live together with a certain amount of mutual understanding – all achieved through conversation.

co-researching R&D in which joint action is based on facilitating the enthusiasms of local people (or stakeholders) and this is autonomous from the research needs and interests of the 'professional' researcher

cybernetics Cybernetics, although often applied to the control of machines, has long been one of the foundations of thought about human communication, its central notion being circularity. Cybernetics 'arises when effectors, say a motor, an engine, our muscles, etc., are connected to a sensory organ which, in turn, acts with its signals upon the effectors. It is this circular organisation which sets cybernetic systems apart from others that are not so organised' (von Foerster, 1992). In first-order cybernetics it was the idea of feedback control which mainly occupied the practitioners (the term was coined by Norbert Weiner in 1948 but originally used by Plato in its Greek meaning of helmsman, or to pilot a craft or 'the art of leading men' (Francois, 1997). In time the question 'what controls the controller' returned to view (Glanville, 1995a,b) giving rise to second-order cybernetics in which the property of circularity became the focus of attention once again.

dialectical relationship As used in second-order epistemology, this relationship continuously exposes and holds together both sides of any distinction and keeps them connected in a recursive way, e.g. the rangelands creating the pastoralists and the pastoralists creating the rangelands. The pastoralist and the rangelands are now seen as a complementary pair: they are distinct but related. The dialectical and recursive process allows us to look at the quality of the relationship as a variable in research.

emotions An emotion, or 'emotioning', is a bodily predisposition to action. Certain characteristics of behaviour can be used to distinguish certain emotions. Psychological theories about emotion have little consistency among themselves but careful observation of our language shows that metaphor is the vehicle by which we reach agreement about such basic states and patterns as emotions.

enthusiasm – theory The word enthusiasm comes from the Greek *en theos*, meaning 'the god within' and is thus distinct from the view that the source of all understanding comes from without. Throughout Western History there has been a tension between whether the primary focus of our understanding comes from nature or from within ourselves.

enthusiasm – biological driving force The emotion or driving force idea of enthusiasm has always been central in psychology. Motivation has been understood as a drive from within that then gets satisfied by whatever you are doing outside. Enthusiasm we conceive of as a drive much like hunger in as much as it is the drive to do. But it is not targeted like hunger. It is the thing that gets us up in the morning to face the world. It acts as a source of meaning which provides the energy that helps us do what we want to do. It gets satisfied in a similar way to the biological drive when we find ourself doing what we wanted to do. We feel the same satisfaction as we do when we are having a meal – satisfying the drive of hunger. And then like any other drive it then quells and returns.

enthusiasm – methodology	Enthusiasm as a methodology is where narrative is employed. The methodology is underpinned by the biological understanding of the drive itself as well as the theoretical principles. It is shaped in a way that does not re-direct a person's energy – it requires a desire to find out where their energy is. This is the initial task in the methodology. To do this requires the right sort of questions: e.g. what do you want to do, why are you a grazier, what is it you get out of this sort of work that is satisfying? To get to the point of having this sort of conversation requires a respect for the individuality of the other and acceptance that whatever they are going to say is valid – based on the notion that it is the god within that person which has to be respected – that is where their energy is. So they can tell their story about where their energy comes from and how they see it expressing itself and what they see as obstacles to its manifestation. It also requires participants to experience being actively listened to.
extension	Agricultural extension as an activity was established in the late nineteenth century in most industrial countries. Over this period it has typically been expressed as a linear process beginning with 'research' and finishing with 'diffusion'. The knowledge is initially generated through research and is then transferred, via extension personnel, to the rural producers. This proposed process has been greatly criticised in recent times but is still the dominant model-in-use.
first-order change	Understands change as taking place in terms of identifiable objects with well-defined properties. The understanding is gained by accepting that there are general rules that apply to situations in terms of those objects and properties. By applying the rules logically to the situation of concern one can draw conclusions about how something has come about and/or what should be done.
first-order data	Such data describes a system as if it was an objective set of operations functioning independently of its historical and social creation.
first-order R&D	Here learning and action are based on the belief of a single reality – a 'real world' – which can be approached and known objectively. It has been characterised by a reliance on a high level of disciplinary knowledge (more recently, multidisciplinary knowledge) and a 'fix' mentality – expose the breakdown and attempt to fix it. In first-order R&D, the problem is clearly defined, the solution is a technological one, and the barriers to adopting the solution are placed fairly and squarely with the end-user community.

first-order tradition	Is characterised by a minimal awareness of how the context actively shapes any experience and especially how the act of observation and participation determines the actual experience. The attitude to knowledge is predominately one of believing in the possibility of an 'objective' knowing of the world. In the rural context, this tradition is characterised by concerned intervention, the definition of clear goals, the naming of the problem, and the proposal of a rational solution.
information	In a technological environment, information is objective, independent of the human mind and can be transmitted without ambiguity. In the world of the observer, 'everything said is said by somebody' (Maturana, 1993) – or second-order cybernetics – information becomes a slippery concept and no effects of communication can be controlled. Information in this view of the world can no longer be said to enter us, it is constituted by us.
institution	The 'rules of the game in which individual strategies compete ...[they] include any form of constraint that human beings devise to shape interaction' (e.g. formal arrangements such as law or informal arrangements such as customary land tenure) (North, 1990).
invitation	A basis for collaboration based on the emotion of mutual acceptance; if an invitation is genuine then one experiences the freedom to accept or decline.
knowledge	In the view of the authors, knowledge does not reflect an objective reality but only an ordering and organisation of the perceptions of the world constituted by our experience. From this perspective knowledge arises in our relationships with 'our' world and is granted by an other even if the other is oneself.
language	Through the use of language, we construct our own reality. We humans have evolved our particular manner of living largely through reliance on the use of language (verbal and non-verbal) as the primary dynamic of any relationship. We are not using language to refer to a means of communicating or transmitting information using symbols or representations of an independent reality.
lineage	A history of descent. No person, tool, machine, book, institution, terrain or ecosystem exists in a current form without descending from something earlier. Like living hereditary lineages, technologies descend and propagate through alliances with other things, by bringing in new forces and assembling new arrangements. Looking at the lineage of things can help tell a story of how the present came to be.
narrative	A way of knowing; it allowed us to find out where people's enthusiasms lie by asking them to tell stories about their lives.

observer	In cognitive science the observer is a technical term which emphasises that we all become observers when we are aware that we exist in a semantic domain created by our operating in language and in the domain of descriptions. By making linguistic distinctions, the observer makes meaning of experience. As observers, we focus on phenomena to be explained.
organisation	The *organisation* of a system is defined as a particular set of relationships, whether static or dynamic, between components which constitute a recognisable whole – a recognisable unity as distinguished by an observer. Organisational relationships have to be maintained to maintain the system – if these change the system either 'dies' or it becomes something else. This is not the common definition of 'organisation' which is often of the more limited and questionable form: 'organisations are groups of individuals bound by some common purpose to achieve objectives'.
participatory rural appraisal	An evolution of rapid rural appraisal (RRA), which draws on a eclectic mix of tools and techniques, in which in an idealised form local people and 'outsider' R&D professionals learn their way to some agreed understanding from which projects or R&D action are formulated.
pastoralist/grazier	Farmer, usually a substantial landholder (on a long-term lease in Australia), who grazes sheep or cattle; a landholder in the pastoral zone or rangelands.
rapid rural appraisal	A systematic but semi-structured study carried out in the field by a multidisciplinary team over a short time used as the starting point for understanding (usually by 'outsiders') a local situation and based on information collected in advance, direct observation and interviews where it is assumed that all relevant questions cannot be identified in advance.
R&D processes	We use R&D as a noun recognising that in everyday use it means something different from its constituent elements of 'research' and 'development'. Whilst it is often used as a synonym for the term 'technology transfer', which refers to the activities of transferring knowledge and practices from one section of the rural community (usually the scientific/marketing/government sector) to another (usually the producers/end-users) with the intention of improving the overall effectiveness of the enterprise, we do not seek to perpetuate this association.
recursiveness	Is the description of a circular relationship such as the one that connects action and experience – all knowing is doing and all doing is knowing – every act of knowing brings forth a world of experience.

responsibility	The capacity for undertaking satisfying action; an expression of autonomy. It is the extent to which stakeholders are enabled to participate in distinguishing a system of interest. We propose as a measure the number of people experiencing an invitation to participate.
second-order change	Change that is so fundamental that the system itself is changed. In order to achieve second-order change it is necessary to step outside of the usual frame of reference and take a meta-perspective (Watzlawick, 1974). First-order change is change within the system, or more of the same.
second-order cybernetics	Is the theory of the observer rather than what is being observed.
second-order data	This data takes as it starting point first-order data such as descriptions of physical, biological, and psychological events with specific reference to the experience (past, present and imagined) of gathering, and working with, the said data.
second-order R&D	Seeks to avoid being either subjective (particular to the individual) or objective (independent of the individual). The objects of our actions and perceptions are not independent of the very actions/perceptions that we make. Problems and solutions are both generated in the conversations that take place between the key stakeholders and do not arise, or exist, outside of such engagements. Second-order R&D is built on the understanding that human beings determine the world that they experience.
second-order tradition	Is characterised by the experience of 'awareness' of being the agent in generating key distinctions (e.g. what is the system under study; what is focused on and what is marginalised; what is the 'problem' and what might be a 'solution') and especially, that the objects and events that we perceive are only knowable through the action of the person perceiving – the 'observer'.
self-reference (self-referential)	The premise underlying 'self-referential' is that as humans, we determine (generate or create) the world that we experience. Second-order understanding implies that all phenomena are self-referential in that they are built, mirror-like, by reference to themselves. The notion of self-reference is in direct contrast to our traditional value of 'being objective' or holding a 'neutral position.'
social construction	Social constructionism has as its understanding that 'all meaningful reality, precisely as meaningful reality, is socially constructed. The chair may exist as a phenomenal object regardless of whether any consciousness is aware of its existence. It exists as a chair, however, only if conscious beings construe it as a chair. As a chair, it too is constructed, sustained and reproduced through social life' (Crotty 1998).

structural coupling	The critical linkage of an organism and its medium (milieu/environment), which could be another organism, is expressed as structural coupling. The autonomous operation of the nervous system – and on the next stage up – the organism as a whole, means that it is capable of changing its own structural dynamics and thus its configuration with its environment. This engagement with another (structural coupling) is dependent on its history of coupling (recursive interactions), with each coupling triggering change which brings about the next possibilities.
structure	The *structure* of a system is defined as the set of current concrete components and relationships through which the organisation of a system is manifest in particular surroundings (see structural coupling and structure-specified systems). This definition is analogous to North's definition of 'institutions' but arises from a different intellectual tradition.
structure-specified (determined) system	A dynamic system (such as a nervous system or, on a larger scale, a human being) in which all the changes that take place are a result of the internal dynamics of the system or are triggered (not directly caused) by interactions with its environment.
system	It is the observer (any person acting purposefully) who, by means of making a number of distinctions (decisions), specifies that a system is a unit distinct from its background. The unit (system) is typically made up of a number of relationships and is given a purpose by the observer and said, again by the observer, to have a function and structure.
systemic action research	The espoused role and action of the researcher is very much part of an interacting ecology of systems. How the researcher perceives the situation is critical to the system being studied. The researcher acknowledges that he/she, in collaboration with others, generates the system.
tradition	In our context, a tradition is a pervasive network of understanding that is largely taken for granted, even unconscious, and which can be looked upon as a 'way of being'. It is the intellectual background within which we interpret and act thus making sense of our experience.
understanding	This is essentially an experience, a form of agreement, a knowledge which arises in our conversations – our living together – and which is not known through properties of something independent of us. Only when we dance in the flow of one another's emotions can we experience understanding.

References

Crotty, M. (1998). *The Foundations of Social Research: Meanings and Perspectives in the Research Process.* Allan & Unwin, Sydney. 55 pp.

Francois, C. ed. (1997). *International Encyclopedia of Systems and Cybernetics.* K.G. Saur, Munchen.

Glanville, R. (1995a). A (Cybernetic) Musing: Control 1. *Cybernetics and Human Knowing*, **3**, 47–50.

Glanville, R. (1995b). A (Cybernetic) Musing: Control 2. *Cybernetics and Human Knowing*, **3**, 43–46.

Maturana, H. (1993). Language and Cognition. Seminar, St Kilda, Victoria, Australia, August 7–9.

Maturana, H.R. and Verden-Zoller G. (1999). *The Biology of Love* (In preparation.)

North, D. (1990). *Institutions, Institutional Change and Economic Performance.* Cambridge University Press, Cambridge.

von Foerster, H. (1992). Ethics and second-order cybernetics. *Cybernetics and Human Knowing*, **1**, 9–19.

Watzlawick, P. (1974). *How Real is Real?* Random House, New York.

Index

action 10, 81, 181
  bodily predisposition to 43
  enthusiasms for 47, 49–50, 151, 157, 172, 179, 215, 217
  as a four stage process 153–4
  enthusiasms for R&D 7
adoption
  failure of 71, 77–8
  lack of technology 6, 61–2
  technology 62, 64
Africa, International Livestock Centre for 12
age, information 44
ageism 118–19
agencies, government 2
agreement 32–3, 36, 45–9
agriculture 123
  NSW Department of 52, 68, 89, 92, 192, 116, 216
  NSW 55, 65, 68, 78–9, 103–6, 108–9, 112, 114, 116, 119–20, 123, 126–7, 129–31, 146–7, 216
  systems 146
Alston, Margaret
  women, exclusion of 118
analysis, textual hermeneutic 77
Andrews, Martin
  interventions, research and technical 18
animals 92–3
  impact of feral 63
appraisal
  critical 215
  participatory rural 146, 186
  rapid rural 27–28, 136, 141–3, 166
approach
  second order 27
  systemic learning 1
  systemic researching 1
assessment 110
autobiography 55

autopoiesis 21

base, knowledge 19
Bateson
  language, metaphorical 35
  patterns 43
behaviour 42–3
biology 13
black boxes 67–9
Board, Western Lands 89
bodies, regulatory 95–8
Bohm, David
  dialogue 45
bores 57–8
  bore-capping 56–7
Bowen, Jill
  Kidman, Sir Sydney 55, 57, 64
braiding 133–5
bush 190

CARR 127, 192–5, 201–4, 206, 216
challenges, political 129–30
change 124, 126, 131
  first-order 3, 124, 136, 214
  second-order 3, 124, 136–7, 215, 217
Clark, Manning
  history 205–6
clubs 111–12
Cobb, John
  knowledge, organisation of 158
cognition 38
  biology of 32
coherence 39–40, 42
  physiological 39
collaboration 27, 133, 137, 141–2
collective 8
colonies 84–5
Commission
  Royal 88, 94–5
  Western Lands 78, 98, 103, 112, 119, 126

communication 19–22, 34, 217
　human 6, 20, 27, 35–6, 38
　theories of 36
community
　rural 1
　sense of 191
complementarity 38
concepts, emotion 43
concern 36
conferencing, search 146
confidence 43, 48–9
　and quality of life 42
　lack of 49
consensus 8
context 161
　anticipated 133
　historical 11–13, 133
　understanding of 55
control 26
conversation 21–2, 37, 45, 121, 150, 153, 158, 186
　network of 129, 150, 181
co-research 103, 106, 161, 163–4, 169, 180–1, 183–5
co-researchers 164, 180–1, 185
co-researching, action 181
Corporation
　Australian Wool 27, 192–4, 196–8
　Wool Research and Development 55, 145, 147, 197, 202
　Corporations, Rural Industry Research 145
correlation, sensory-effector 40
coupling, structural 22, 40–2, 70
creation, technology 18
CSIRO 98
cybernetics 34
　first order 34
　second order 25, 33–4, 36

Darwin, Charles
　animals, emotions of 43
data 133
　first order 214–15
　second order 133, 214
deep-ending 107–8
demands, political 121
Department, the 110–11

dependence 58–61
dependency 126–7
determined, structure 39
dialectic 25
dialectical 25
dialogue 45
diffusion, technology 65
discourses 61
discussion 45
diversity 164
drought 95
dualism, duality and 24
duality 8, 24

ecology, social 146
education, agricultural 27
effector 40
Ellis & Swift
　ecosystems, African pastoral 17
emotioning 43, 47
emotions 36, 42–5, 47–8, 153
　and language 36–7
engagement 37–40
enthusiasm 7–8, 28, 47–8, 134, 136–7, 142–3, 145–8, 153–8, 172, 179, 215
　as a biological driving force 143
　as an environment for triggering 156
　as a methodology 143–5, 154–5
　as a theory 143, 146–7
　notion of 179
epistemology, cybernetic 25
equation, extension 19
erosion, soil 96
ethics 2, 48, 58
evaluation 151
　critical 184
executives, senior 106–7
experience 1, 6, 37, 44, 142, 157, 164–6, 168–9, 181–4
experiments, development 18
explanation 36–7, 64
　scientific 156–8, 219–22
explorers 83–4
extending 72–3
extension 11, 19, 103–4, 106, 109, 114, 116, 120, 124–5, 131, 146

extension (*cont.*)
  agricultural 1
  rural 1

facts 67–8, 80–2, 101
family, relationships with 115
farmers 2, 28, 32, 46, 123, 136, 141–2
Fell & Russell
  development, second order research and 47
fiction 55
Fortmann, Louise
  Botswana, rangeland use in 15
Foucault, Michel
  power, discourse 77

Gap, Fowlers 5
Galileo
  technology 64
government 83, 124, 128
Grazfeed 93
graziers/pastoralists 3, 7, 18, 25, 28–9, 52–3, 58, 62–3, 66, 69, 73, 122, 133, 147, 149–151, 154–5, 162–5, 167–9, 172, 176, 178–81, 184, 186, 193–4, 205
  and co-researching 161–3
  conversations with 150, 154
  dialectic, pastoralist/rangeland 25
  experience, reflection on 73
  father, Ray's 66
  families of 133–4
  goals of 124–5
  innovations, introducing technological 90
  lives, technologies which transformed 71
  manager, grazier as 87
  numbers, reducing grazier 99
  overgrazing, overstocking and 18
  overstocking 63
  pasture, technologies of the 96–9
  relationships with 112–13, 123–4
  researching, experiences of ways of 7
  returns, failure to send in annual stock 88
  SSIs with 141, 164–5, 168–9, 148
  stories 58, 61, 70
  system, developing a solar power 60
  technologies, animal 94–3
  technology, designed-in dependence on 58
  technology, failure to adopt 52–3, 55
  term 88
  the story of 189–204
  view, mainstream 16
grounding, contextual 10, 16, 161, 164
Group, UK Farming and Wildlife Advisory 117
groups 148–9, 155
  farmer 2
  user initiated R & D 134
gyrocopter 53

helplessness, learned 42
history 82
  experiential 6
Hooker, Cliff
  amplifier, technology as a selective 65

Ihde, Don
  artefact 70–1
  design 70–1
imagery, satellite 87
impacts, technological 65
incident, critical 142, 146, 149–50, 154
independence 59
industry, wool 194–5, 198, 201
innovation 57
  grazier 89–90
  technological 78, 91, 92
inquiry, streams of 6
installation, grid power 59–60
institution 105, 126, 128, 131
interest, systems of 6
interviews 60
  in-depth 161–2, 165–7
  semi-structured 28, 79, 103–4, 106, 127, 133, 147, 150, 164–9, 171–2, 183–4, 186
isolation 191
Ison

ageism, gender and 118
Ison & Blackmore
 complexity 6
issues 36

Jiggins, Janice
 extension 11
Johnson, Robert
 enthusiasm 142
 ritual 153
joints, universal 59

Kersten
 rangelands, researchers in semi-arid 124
Kelly
 position, constructivist 162
knowing, the biological basis of 20–1
knowledge 1, 6, 18, 22, 29
 applying 18
 instructing with 20
 new 12
 transferable 19–20
Kövecses
 concepts, emotion 43
Krippendorff
 communication, metaphors of 35, 44

land 88, 90–1, 94, 97–8, 100
Lands, Department of 89
language 35–7, 44–5
 and emotions 36
 biology of 35–6
 metaphorical 44
Latour, Bruno
 black boxes 67–70
 engineers, scientists and 10
 model, diffusion of innovation 67–70
Lawson, Henry
 drought 70–1
Leach & Mearns
 Africa 16
learning, experiential 181
Le Houerous
 Sahel, African 14
 tradition, first order 14

Leunig, Michael
 living, track laid down in 14
life
 quality of 41
 technologies of animal 92–3, 94–5
lineage 81–2, 98
 technological 73, 77–8, 86–7, 92–3, 131
love 207

management 97–8, 100–1, 117–18
 range 12–13, 17
managers
 middle 106–7, 111, 120, 121–22
 senior 111, 116–17
map 83, 84–5, 87, 100
 survey map 84, 86
Margerum & Born
 organisation 128
Maturana, Humberto 7, 14, 23, 36–8, 43, 47–8, 70, 73, 150, 153, 155, 156–8, 169, 206–7
 change, triggering 169
 conversations 153, 158
 coupling, structural 14, 70
 double look 38, 40
 emotioning 43
 emotions 47–8
 explanation 157
 human, what it is to be 37
 invitation 150–1
 phenomenon 156
 physiology, behaviour and 36
 research, social 7
 science 23
 technologies 73
 technology, science and 206–7
Maturana & Varela
 autopoiesis 35
 cognition 44
 conversation, network of 181
 explanation, biological 162
 phenomenon 156
 reflection 137
 theories, cognitive 70
McClintock
 transfer, technology of (TOT) 117

meaning   27, 45
  shared   21
  as a relational phenomenon   27
meaningful   165
measures, quantitative   97
meetings   151
metaphor   1, 9, 139, 141–2, 144, 148, 157, 218
  conduit   20
  extension   116–17
  for human communication   35
  hypodermic   20
  organising   142, 157
  theory of   35
mills   198–200
  Wagga Wagga   198–9, 200, 202
mirroring   148, 169, 173, 175–6
Mitchell, Thomas
  surveying-exploration   83–4
model   128
  agricultural R & D   64
  diffusion   67–8
  diffusion of innovations   65, 67–8, 70, 116
  ecosystem   99, 101
  farmer-led   19
  role   108–9
monitoring   98–9, 101
Morgan, Gareth
  explanation   157

narrative   137, 143–4, 165–7
nesting   177–8, 187
networks
  building   65
  people in   70
North, Douglas   105
NSW, Western Division of   4, 52–3, 55, 57–8, 63, 73, 80, 82, 84, 86, 88–9, 91–101, 103, 111–12, 119, 124 145, 161
objectivity   10, 13, 15, 26, 35
observation   181
observer   26
officers, extension   107–14, 116–19, 121, 124–5, 131
organisation   6, 55, 78, 103–5, 107, 109, 119–21, 123–4, 126, 128–30

service   103, 106, 108, 123, 125
  NSW State Government Research and Extension   78
overstocking   95
OZPIG   130

packages, management   93
participants   176–7, 179–80
pasture
  technology of   94–8
patterns   73, 77, 167
  hermeneutic   77
  historical   73–4, 77
  meaning-making   167
  on the ground   167
people   81
perception   70, 146
personnel, extension   106–7
perspective, diffusionist   68
phenomenon   156–7
planning   181
poetry   55
poor, information   136, 141–2, 157
posters, thematic   170–1
power   26, 78, 80, 82, 86, 89–92, 100
  object of   91–2
  relations of   82, 86, 88–91, 94, 97, 100
practice   162
practice, extension   110
  and gender   117–18
prejudices   15
pre-understandings   165, 168
problem   14–15, 25, 62, 77–8, 95, 97–8
problems, system-generated   216
production, technologies of 82–3
production   94–5, 100–1
productive   92
productivity   91, 93, 95
procedure, sampling   77
professor   32, 46
project
  Systemic Action Research   1–2
  R & D   5
  research   52, 55, 68, 145, 192, 205

question
  descriptive 164
  raw data of experiences 164

rabbits 95
rangelands 6, 13, 17, 24–5, 28, 61, 78, 95, 99, 104, 120, 131, 161
  organisations operating within 103–32
  relationships with the 115
  semi-arid 3, 53, 69, 73–4, 80, 115, 124, 133, 147, 208
RANGEPACK 93
realities, multiple 20, 163
reality, fixed 20
recognition 110
reflection 42, 137, 181
  critical 7, 137, 154, 156–7, 165, 181–2, 184
regime, coherent 40
relationship 111–12, 114–15, 123, 131
  research as a dialectical 25
  research–development 24
  first order and second order R & D 26
repairs 59
research 7, 95–6, 137, 146, 148, 161, 174–6, 186, 189, 192–3, 195–6, 201–2
  action 5
  artificial intelligence 22–3
  descriptive narrative 165
  farming systems 6
  qualitative 78
  scientific 100
  social 7
  systemic action 5, 7, 183
responsibility 10, 27, 217
rich, information 136, 141–2, 157
ritual 153
rivalries 119–20
river
  Murray 55
  Darling 55
Rodgers, Everett
  innovations, diffusion of 65–7
Röling, Neils
  interface, research–technology transfer 22
Russell & Ison
  development, second-order research and 47
rural research and development (R & D) 1–3, 5–7, 9–12, 37, 64, 70, 73, 134, 145, 189–218
  and driving ideas 2
  and structural needs 120–1
  institutionalised 70–1, 73
  first order 10, 12–13, 24, 26, 158
  natural resource 11
  rural 7, 12
  second order 10, 12–13, 23–4, 28, 47, 110, 133–4, 137, 158, 184

Sanford, Stephen
  development, pastoral 18
  view, mainstream 16
science 48, 64, 91
  contextual 26
  first order 24
  second order 24
  range 16–17
Scoones, Ian
  Africa 11
self-reference 8, 25
self-referential 25
sensory 40
sensory-effector 40
Service, Soil Conservation 78, 92, 96, 98, 108, 119
session
  plenary 177–8
  small-group 176
Shannon & Weaver
  communication, human 20
  transfer, information 67
SHRUBKILL 93
Sless, David
  communication 20
society 1
solution 15–16, 125
staff, relationships with 113–14
stakeholders 125, 127
statement, action 180
story 7, 16, 18, 80–2, 99–100, 130, 133, 144–7, 165–7

story (*cont.*)
  David's   137–9
  Ray's   139–41
  rich   7
  unfinished   7
storytelling   165–6
stress   42
structure   37–8, 195–6, 123, 131
Sturt, Charles
  Division, Western   83
sub-themes   107
supplies, water   57–8
surveillance, technologies of survey and   82–3
survey   85–8, 100
  government   83–4, 87
  technology of the   86–7
surveying-exploration   83
Swift
  organisations, institutions and   104–5
Sydney
  University of   146
  University of Western   146
Sylla
  organisations, African pastoral   128
system   3, 26, 105, 125–6, 208–9, 217
  action   208
  Agricultural and Information   37
  autopoietic   37–8
  biology of observing   33
  closed   10, 15, 37
  decision support   93
  equilibrial   15
  expert   93
  first order R & D   209
  global R & D   10–11
  human   20
  information transfer   21
  knowledge and information   22
  learning   208
  living   37
  nervous   21, 38, 40
  non-equilibrial   16
  open   10
  physiological   39
  researching   208
  R & D   208–13
  self-regulating   38
  structure specified   20
  structure determined   21–2

talking   55
team   151
technology   6, 55–7, 62, 64–5, 68–70, 812, 92–5, 98, 100
  administrative   100
  as a selective amplifier   65
  mechanised   53
  new   12
  survey and surveillance   83
  transfer of (TOT)   116–17, 131
  transfer or knowledge of   64–5
themes   106–7, 166, 171
thematic   166–7
there, out   18, 21, 23
theodolite   83
theory   162
  communication   34
things   81–2
tradition   1, 3, 5–7, 14
  farmer first   6, 28
  first order   3, 5–7, 9, 13–16, 18, 74
  second order   5, 7–9
trajectories, technological   7
transfer
  information   19–20, 34, 47, 67
  innovations model of technology   65–6
  technology   6, 11, 18, 47, 55, 64, 79, 145, 192
  technology of   6, 55, 64–5, 116–17, 175
translation, model of   68–70
triggering   142, 158
triggers   38, 142
Trough, Finlayson   56–7
truths   80–1
Tucker   143, 155, 158
  enthusiasm, transience of   155
  science   158

understanding   29, 32, 33, 36, 44–6, 48–9, 166
  frameworks of   15
  human   6

models of   3, 14
traditions of   5, 16–18
unity   8

Varela   38
  system, nervous   38
view
  individual world   162
  mainstream   16, 18
Von Foerster, Heinz
  confidence   49
  cybernetics, second order   34–5
  shift, historical   25
  tradition, first and second order   5

Wadsworth, Yoland
  graziers, targetting of   68–9
Watzlawick
  change, second order   136

weeds, woody   53–4, 62, 64, 123, 170
  burning of   62
  technologies for control of   62, 71
Winograd, Terry
  intelligence   23
Winograd & Flores
  intelligence   23
  rationality   13
  understanding   47
women   28, 172, 191
wool   194–200
workshop   79, 134, 139, 172–3, 177–9, 193–4
world
  external   18
  real   10, 21
Wright, Susan
  Britain   79